Nonmetallic Materials and Composites at Low Temperatures 2

CRYOGENIC MATERIALS SERIES

Nonmetallic Materials and Composites at Low Temperatures
Edited by A. F. Clark, R. P. Reed, and G. Hartwig

Filamentary A15 Superconductors
Edited by Masaki Suenaga and A. F. Clark

Nonmetallic Materials and Composites at Low Temperatures 2
Edited by Günther Hartwig and David Evans

A Continuation Order Plan is available for this series. A continuation order will bring delivery of each new volume immediately upon publication. Volumes are billed only upon actual shipment. For further information please contact the publisher.

Nonmetallic Materials and Composites at Low Temperatures 2

Edited by

Günther Hartwig

Nuclear Research Center
Institute for Technical Physics
Karlsruhe, Federal Republic of Germany

and

David Evans

Rutherford Laboratory
Oxfordshire, England

Plenum Press • New York and London

Library of Congress Cataloging in Publication Data

Main entry under title:

Nonmetallic materials and composites at low temperatures 2.

(Cryogenic materials series)
"Proceedings of the second ICMC Symposium on Nonmetallic Materials and Composites at Low Temperatures, held August 4–5, 1980, in Geneva, Switzerland"
Includes bibliographical references and index.
1. Nonmetallic materials – Thermal properties – Congresses. 2. Composite materials – Thermal properties – Congresses. 3. Materials at low temperatures – Congresses. I. Hartwig, Günther. II. Evans, David, 1940 Oct. 8-   . III. ICMC Symposium on Nonmetallic Materials and Composites at Low Temperatures (2nd: 1980: Geneva, Switzerland) IV. Series.
TA418.95.N66   1982   620.1′1216   82-367
ISBN 0-306-40894-5   AACR2

Proceedings of the Second ICMC Symposium on Nonmetallic Materials
and Composites at Low Temperatures, held August 4 – 5, 1980, in
Geneva, Switzerland

© 1982 Plenum Press, New York
A Division of Plenum Publishing Corporation
233 Spring Street, New York, N.Y. 10013

PREFACE

This, the second special topical conference on the properties of Non-Metallic Materials at Low Temperatures, was sponsored by the International Cryogenic Materials Conference Board.

The potential for plastics materials in the field of cryogenics is vast and as yet only partly explored. In addition, many other materials, which qualify for the title non-metallic but are not 'plastics', have numerous possible outlets in low temperature technology. This conference aimed at providing a forum, whereby specialists from Industry, the Universities and from Government sponsored Institutions could assemble to discuss the extent of our current knowledge. As it transpired, the meeting was also to highlight the considerable gaps that still exist in our fundamental understanding of the low temperature behaviour of these materials. On this theme, during the course of the conference, a reference was made to an almost forgotten quotation by Lord Kelvin, who said:

"When you cannot measure what you are speaking about, when you cannot express in numbers, your knowledge is of a meagre and unsatisfactory kind; it may be the beginning of knowledge, but you have scarcely in your thoughts advanced to the stage of a science, whatever the matter be."

This simple statement sums up the aims, objectives and hopefully the achievements of this conference. To discuss and disseminate the current knowledge on non-metallic materials in order that realistic predictions of in-service performance may be made.

When workers involved in assessing fundamental properties can meet with and discuss the problems of the Engineer who must use these materials in realistic applications, then a healthy and vigorous atmosphere is created. Confidence in the use of these materials is growing, as is the awareness of the wide range of materials available but with this growth grows the danger of applying the wrong material in the wrong application at the wrong time. Non-metallic materials and the cryogenic community are in a period of consolidation when knowledge and experience are growing side by side. This delicately balanced relationship could be damaged by an expensive or dangerous in-service failure and cryogenics as a whole would suffer from a loss of confidence in non-metallics. For these reasons the International Cryogenic Materials Conference Board feels that special topical conferences of this type are valuable by helping to ensure that piece by piece and year by year the application of non-metallic materials to the low temperature environment advances to the stage of a science.

If we paraphrase Lord Kelvin a little further, then it is necessary to describe the properties and behaviour of non-metallics at low temperatures in terms of numbers in order to accurately predict their behaviour. In this science we, as researchers and users, have a long way to go to catch up with the head-start enjoyed by metallic materials. In addition, we have to cope with the complications of formulation and manufacturing variables which is a problem overcome by standardisation many years ago in the field of the true metallics. The cryogenic community stands poised, ready for large scale applications of low temperatures in a series of commercially viable and beneficial projects. Those of us involved with non-metallic materials realise that although we don't yet know all the answers, at least we have asked ourselves the questions and expressed our knowledge in at least some of the numbers we need.

The editors wish to express their sincere appreciation for the efforts of Mrs Janice Brown who typed the greater part of this volume.

David Evans
Gunther Hartwig

CONTENTS

PANEL DISCUSSION

THERMAL PROPERTIES OF CRYSTALLINE POLYMERS AT LOW TEMPERATURES

I Engeln and M Meissner

Institut fur Nichtmetallische Werkstoffe and Institut
fur Festkorperphysik
Technische Universitat Berlin, D-1000 Berlin 12, W Germany

INTRODUCTION

In comparison with inorganic solids there have been few studies
of the low temperature thermal properties of polymers, even though
these properties are of considerable interest in scientific research
and cryogenic technology. From the three important thermal
quantities, heat capacity Cp, thermal conductivity K and thermal
expansion α, the heat capacities of polymers have been most widely
studied[1]. Most of the studies on thermal conductivity of polymers
are confined to the range between liquid nitrogen and room
temperature. Recently, however, the situation has improved and
thermal conductivity data[2] for polymers are now available in the
temperature range 0.1 to 300K. In contrast, there is a lack of data
on thermal expansion measurements at low temperatures. On the other
hand there has been a growing theoretical interest in the lattice
dynamics of polymers representing a rather extreme form of anisotropy
in the crystalline state[3].

Over the last ten years low-temperature measurements of Cp and
K have been made on wide range of amorphous solids, including a
number of amorphous polymers. For all of them similar features
have been observed: between 5 and 15K the heat capacity is showing
an excessive heat in addition to the Debye-contribution and the
thermal conductivity is almost independent of T, yielding K-values
between 10^{-3} to 10^{-4} W/cm.K; below 1K the heat capacity varies as
Cp $\sim T^1$ and the thermal conductivity is approximately proportional to
T^2 (see Figures 1 and 2). These "anomalous" properties of all
amorphous solids do not seem to be seriously influenced by structures[4], i.e. inorganic solids, compounds and polymers. It has been
proposed [5,] that the thermal anomalies in glasses arise from a

1

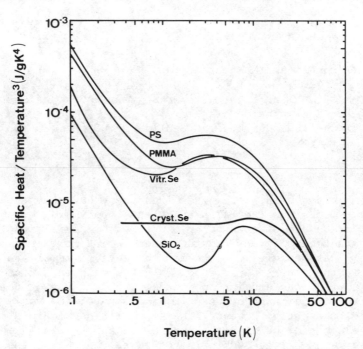

Figure 1. Specific heat of inorganic glasses, SiO_2 (Zeller and
Pohl[35]), and Se (Lasjaunias et al[36], Meissner and Wobig[23], and
of polymer glasses, PMMA and PS (Stephens et al[37], Choy et al[15]
and Wunderlich and Baur[1]), at low temperatures,plotted C_p/T^3
versus T. For comparison trigonal (polycrystalline) selenium
(Lasjaunias et al[36], Meissner and Wobig[23], is showing a Debye
T^3 law down to 0.3K.

distribution of localized two-level tunnelling states characteristic
of the amorphous state. Despite the existence of a number of other
theories the tunnelling model seems to be the most successful in
explaining the temperature dependence of heat capacity, thermal
conductivity, ultrasound propagation[6] and the Gruneisen-Parameter[7].

Most polymers are either amorphous or semi-crystalline and the
thermal properties are strongly influenced by the degree of crystallinity. Thus, it may be of considerable interest to research the
thermal properties of crystalline polymers. Unfortunately there have
been only a few studies on heat capacity, thermal conductivity and
thermal expansion of crystalline polymers. Low temperature
measurements have been reported of heat capacity[8] and thermal
conductivity for highly crystalline polyethylene, with a degree of
crystallinity X≈0.8. The first measurement of the thermal expansion
of a polymer crystal (polyoxymethylene) was reported in 1976[9], but
there was a marked lack of perfection in the crystal and it was
possible to make only a rough estimate of the thermal properties of
the crystalline phase. In contrast with the experimental side
there has been a growing interest in the lattice dynamics and
associated physical properties of polymers. Baur[10] has described
bending and stretching modes of the polymer lattice due to the
stiffness of the polymer backbone. As these modes contribute to the
vibrational density of states $g(\omega)$ one may observe a characteristic
temperature dependence of the heat capacity of a polymer crystal.
In order to explain the anisotropic thermal expansion of polymer
crystals, a number of models has been proposed[11] [12] [13] [14] [3].

The results are commonly given in terms of the Gruneisen
parameter γ, which relates the thermal and mechanical properties
of a solid. Due to the strong anharmonicity of the low frequency
acoustic modes in a polymer chain lattice one can expect bulk values
at T = 0 K between γb = 3.16[14] and γb = 5.25[12]. As Gibbons has
shown,[3] there is additionally a marked difference in γ-values calculated from interchain ($\gamma\perp$) and intrachain ($\gamma\parallel$) forces. Low
frequency transverse modes propagating along the chains will have a
low $\gamma\parallel$, whereas perpendicular to the chains $\gamma\perp \simeq \gamma b$.

The purpose of this paper is to discuss some recent experimental results on the thermal properties of crystalline polymers,
i.e. heat capacity, thermal diffusivity and conductivity, and also
the thermal expansion and Gruneisen-function of semicrystalline
polyethylene (PE) and single crystalline polydiacetylene (PTS)
in the temperature range from 3 to 300K. For PE (X = 0.44 to 0.98)
our results complete earlier measurements in Cp[8] and K[15], showing
that for differing degrees of crystallinity the thermal properties
can be described by a two-phase model. Comparing the results on
nearly polycrystalline PE with those for the PTS-crystal there are
marked new effects at low temperatures (T≲100K): the heat capacity
shows a temperature dependence[16] as expected for a polymer chain

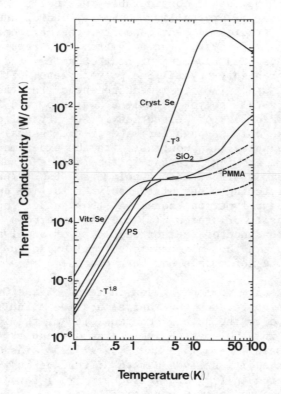

Figure 2.   Thermal conductivity of inorganic glasses, SiO$_2$ and
Se (Zeller and Pohl[35]), and of polymer glasses, PMMA and PS
Stephens et al[37] and Hager[38]at low temperatures. For comparison trigonal (polycrystalline selenium is showing a Debye T^3
law down to 2.5K (Meissner and Wobig[23])

lattice by Baur[17]. The thermal conductivity is characterized by a nearly constant anisotropy $K||/K\perp$ = 15 and, in addition, shows a temperature dependence that deviates strongly from that for crystalline solids. Measurements of the thermal expansion have been performed for both polymers down to 100K and for two semi-crystalline PE-samples (X = 0.59 and 0.77) down to liquid helium temperature. Analysing these measurements in terms of the Gruneisen-parameter the theoretical predictions of $\gamma_b \simeq 4$ are supported.

In order to sum up the thermal properties of polymers the paper is organized as follows. In section 2 the heat capacity of semi-crystalline PE with its dependences from crystallinity and temp-erature is described. Some related effects arising from the model of the ideal polymer lattice are discussed with recent Cp data on a PTS-single crystal. In section 3 thermal conductivity measure-ments on PE and PTS are compared, together with the results presented by Greig and Hardy[18]. In section 4, our first results of thermal expansion measurements on PE and PTS are presented. Although the low temperature data on semi-crystalline PE supports current theories, it would seem that the continuation of thermal expansion measurements on polycrystalline PE and single crystalline PTS down to 4K is essential.

HEAT CAPACITY

The heat capacity of a polymer material is influenced by its degree of crystallinity X. Thus, we start with the temperature dependence of the heat capacity of a polymer crystal, polydiacetylene toluene-sulphonate (PTS), which can be produced yielding nearly perfect single crystals[19]. In Figure 3 the specific heat at constant volume $C_v(T)$ of PTS is shown between 2.5 and 300K[16]. Three tempera-ture regions can be distinguished: below 7K Debye's T^3 law is observed., in the temperature range from $\sim$15K to $\sim$50K and from $\sim$50K to $\sim$200K $C_v(T)$ varies as $T^3 + T^{3/2}$ and $T^1 + T^{1/2}$, respectively. These three temperature regions can be explained by the existence of bending (b) and stretching (s) phonons of the polymer backbone.

The importance of b-phonons contributing to C_v of a polymer crystal has been pointed out by Baur[17], showing that polymer lattice vibrations coupled with the bending force constant of a polymer chain are characterized by anomalous dispersion relations $\omega(k) \sim K^2$, where ω is the phonon frequency and k the wave number. For very small phonon frequencies (T$\to$ 0 K) $\omega(k)$ becomes linear, due to the inter-chain interactions in a polymer lattice. In case of s-phonons the linear dispersion relation (for small, but not too small frequencies) holds independently of the direction of propagation. Even though the planar conjugated backbone of the PTS chain differs markedly from the linear chains of the Stockmayer-Hecht model[20], it is assumed that the above behaviour is not principally affected by this.

Using a continuum approximation, the phonon density of states $g(\omega)$ has been calculated[16]. At higher frequencies (one dimensional approximation) $g(\omega) \sim \omega^0$ and $\sim \omega^{-1/2}$ for s- and b-phonons, respectively, are found yielding $C_v \sim T^1$ and $C_v^b \sim T^{1/2}$. At lower frequencies the phonon wavelength becomes greater than the chain distance in the polymer lattice; thus, in the three dimensional approximation one yields $g(\omega)^s \sim \omega^2$ and $g(\omega)^b \sim \omega^{1/2}$. At sufficiently low temperatures the temperature dependence of the specific heat should change into $C_v^s \sim T^3$ for s-phonons and $C_v^b \sim T^{3/2}$ for b-phonons. As $\omega(k)$ for all phonons becomes linear at very small frequencies the polymer lattice is showing Debye's $g(\omega) \sim \omega^2$ yielding $C_v \sim T^3$. Constructing a density of states as outlined above the heat capacity of PTS has been calculated. The good agreement between theory and experimental result is shown by Figure 3.

Considering semi-crystalline polyethylene, four isotropic poly-ethylenes Lupolen* 1840 D, $X = .44$ (PE 44), Lupolen 4261 A, $X = .62$ (PE 62), Lupolen 6011 DX, $X = .77$ (PE 77), and Lupolen 6011 L, $X = .98$ (PE 98) were investigated. The first three samples are known to have a lamellae thickness[21] between about 15 and 25 nm. PE 98 was pressure crystallized and from electron scan microscopy we estimate its lamellae thickness to be about 440 nm. From this we conclude the amorphous layer between crystal lamellae to be 4 nm or less.

Specific heat results from 3 to 300K are shown in Figure 4. In addition, extrapolated C_p - data of completely amorphous PE ($X = 0$) are shown[1] between 3 and 200K. In the temperature region 3 to 230K the data for PE 98 is in good agreement with the extrapolated values for 100% crystalline PE. Increasing the amorphous fraction does not affect the specific heat between 50 and 200K, which is the temperature region where the one-dimensional phonon spectrum, characteristic for polymer solids, dominates. Below 50K a dependence of the heat capacity on crystallinity is observed.[22]

Figure 5 illustrates the excessive heat contibution to the crystalline heat capacity by plotting the low temperature data as C_p/T^3 versus T. For PE 98 a Debye T^3 law is observed below $\sim$10K, and for PE samples with lower crystallinity (PE 77, PE 62, PE 44, and PE 0) a hump with a maximum at $\sim$5K is shown. It is assumed that the origin of the excess heat capacity is characteristic of the amorphous state and arises from a small number of low frequency transverse acoustic modes or non-acoustic modes of low frequency[4]. The appearance of non-Debye low frequency modes related to the amorphous structure is supported by testing the low temperature specific heat of PE as a function of crystallinity. In Figure 6 the low temperature C_p data are shown to depend linearly on crystallinity for temperatures between 2.5 and 25K, which means that

---

* Lupolen is a trademark for PE made by BASF AG, Ludwigshafen, Germany

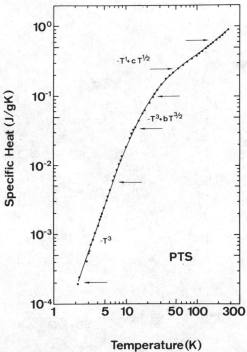

Figure 3. Specific heat of a polydiacetylene (PTS) single crystal between 2.5 and 300K. The straight line is calculated from a chain lattice model for a polymer crystal (Engeln and Meissner[16]. The arrows are indicating the region where the temperature dependences are valid.

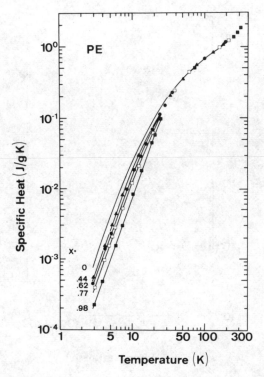

Figure 4. Specific heat C_p of semi-crystalline polyethylene
(PE) with varying degree of crystallinity X. The straight
lines below 20K were calculated from Tarasov-model with one
additional Einstein-term according to eq. (2). Data for X = 0
have been extrapolated from Wunderlich and Baur[1].

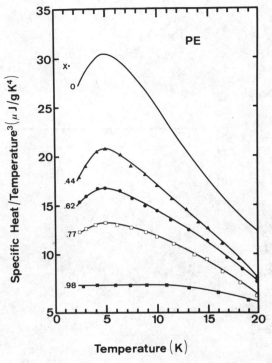

Figure 5. Low temperature specific heat of semi-crystalline polyethylene, plotted C_p/T^3 versus T. The straight lines are according to eq. (2).

TABLE 1. CHARACTERISTIC TEMPERATURES θ_1, θ_3, θ_E (IN K) AND RELATIVE CONTRIBUTION OF EINSTEIN-MODES N_E/N (IN %) USED TO FIT EQ. (2) TO LOW TEMPER-ATURE C_p DATA OF PE (SEE FIGURES 4 AND 5).

X	θ_1	θ_3	θ_E	N_E/N
0		88	23.4	0.37
.44		109	25.4	0.35
.62	871	111	23.8	0.14
.77		122	24.0	0.10
.98		142	–	–

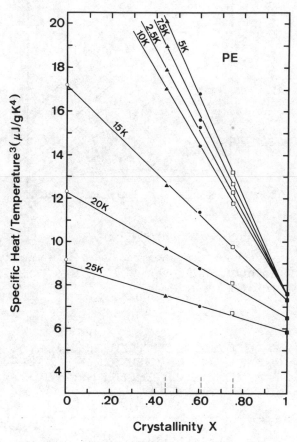

Figure 6. Low temperature specific heat C_p/T^3 of semi-crystalline polyethylene as a function of crystallinity. The open circles refer to data from Wunderlich and Baur[1] for X = 0.

with decreasing degree of crystallinity a linearly increasing number
of excess modes must be added to the phonon spectrum of crystalline
polyethylene.

Thus, we have tried to analyse the low temperature C_p data of PE
in terms of a unique phonon density of states $g(\omega)$. In a simple
approximation to the polymer lattice model we used a Tarasov-model,
which has often been applied to solids with one dimensional chain-
like structure. At high frequencies $g(\omega)$ is described by a constant
density of states with a lower characteristic frequency ω_3 and a
highest value ω_1, whereas in the region of acoustic phonons ($0<\omega<\omega_3$),
$g(\omega) \sim \omega^2$ is used. For semi-crystalline PE ($X< 1$) Einstein modes
with characteristic frequencies ω_E and number of modes N_E are
introduced in order to describe the low frequency modes characteristic
of the amorphous matrix. From the above defined $g(\omega)$ one is able to
calculate specific heat[23].

$$\frac{C_v}{3N \cdot k_B} = (1 - \frac{N_E}{N}) \left\{ D_1\left(\frac{\theta_1}{T}\right) - \frac{\theta_3}{\theta_1}\left[D_1\left(\frac{\theta_3}{T}\right) - D_3\left(\frac{\theta_3}{T}\right)\right]\right\} + \frac{N_E}{N} P\left(\frac{\theta_E}{T}\right) \quad (1)$$

where $D_{1,3}$, P represent the Debye and Einstein function, respectively,
$\theta = n\omega/k_B$ are the characteristic temperatures and N, k_B,
have their usual meanings. Assuming that the maximum characteristic
frequency ω_1 is of the same value for all crystallinities, we
obtained from a fit of the C_p data between 80 and 200K a value
$\theta_1 = 871K$. Now we were able to fit all low temperature C_p data with
3 parameters A, B, θ_E by means of best fit procedure according to

$$(T \leqslant 20K) \quad C_v = A \cdot T^3 + B \cdot P\left(\frac{\theta_E}{T} \right) \quad\quad (2)$$

An excellent agreement could be achieved between calculated and
experimental data as shown in Figure 5 by the straight lines. In
table 1, the variation of the model parameters θ_3, N_E (calculated
from A and B, respectively) and θ_E with crystallinity are shown.
θ_3 show the softening of the PE "lattice" with decreasing
crystallinity and, as a consequence, the one dimensional continuum
$g(\omega) = $ const. moves to a lower frequency. The additional Einstein
modes are located at approximately 24K, independently of X, and the
number of modes N_E is increasing linearly with X (for the branched
PE 44 the value does not fit the linear dependence).

To find physical evidence for the suggested low frequency modes
we studied the FIR absorption of the PE samples[24]. Unfortunately,
reliable data of the optical constants could not be obtained below
20 cm^{-1} ($\sim$30K characteristic temperature). So far, only FIR
studies on glasses (SiO$_2$, PMMA) were successful in finding relevant
absorption at very low frequencies[4].

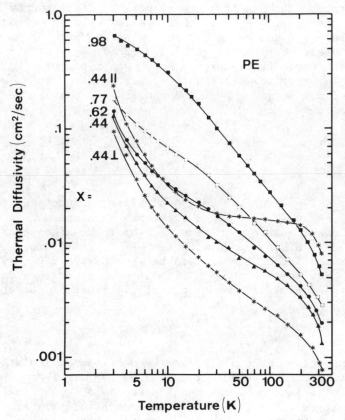

Figure 7. Thermal diffusivity of semi-crystalline polyethylene
with varying degree of crystallinity X. A low density sample
has been extruded (draw ratio 6), and was measured parallel
(.44‖) and perpendicular (.44 ⊥) to the draw direction.

THERMAL CONDUCTIVITY

To get an insight into heat transport properties, not only is
the knowledge of the phonon spectrum at a given temperature important,
but also the phonon-phonon interaction and phonon-structure
scattering in the material have to be taken into consideration.

In the relaxation time approximation, the thermal conductivity
K is given by the density of states $g(\omega)$, specific heat $c_i(\omega)$,
velocity $v_i(\omega)$ and relaxation length $l_i(\omega) = v_i(\omega) \cdot \tau_i(\omega)$
(τ = relaxation time) of phonons

$$K = \frac{1}{3} \sum_i \int_\omega g(\omega)\, c_i(\omega)\, v_i(\omega)\, l_i(\omega)\, d\omega , \qquad (3)$$

where integration for each phonon branch i has to be carried out
over all heat carrying phonon frequencies ω. In a simplistic view
the dominant phonon approximation yields

$$K = \frac{1}{3}\, C_D\, \bar{v}\, \bar{l} , \qquad (4)$$

with $\bar{l}$ as the mean free path, $\bar{v}$ as the mean sound velocity, and C_D
as the specific heat of the heat carrying phonons.

Experimentally, the thermal conductivity can be obtained from
heat pulse experiments yielding the thermal diffusivity a :

$$K = \rho \cdot C_p \cdot a , \qquad (5)$$

where ρ is the density of the sample[25]. Thermal diffusivity
measurements have been carried out on the isotropic PE samples
PE 44, PE 62, PE 77, and PE 98. Results are shown in Figure 7.
One sample Lupolen 1810 H having a volume crystallinity of X = 0.44
has been extruded at room temperature with a draw ratio of 6 and
diffusivity has been measured parallel (.44$\parallel$) and perpendicular
(.44$\perp$) to the draw direction, resulting in a $a_\parallel/a_\perp \simeq 10$ at 200K and
$a_\parallel/a_\perp \simeq 3$ at 3K.

Combining eqs. (4) and (5) the phonon mean free path $\bar{l}$ can be
calculated. This has been done using appropriate C_p/C_D and $\bar{v}$ values
extracted from specific heat data. The influence of the degree of
crystallinity X on the temperature dependence of $\bar{l}$ can be clearly
seen in Figure 8. The temperature behaviour of PE 44, which has the
highest amorphous fraction, already shows some similarities[2] to
$\bar{l}(T)$ of completely amorphous materials. Only little variation of $\bar{l}$
with temperature is seen between 50 and 250K, due to the amorphous
structure. In this temperature region PE 76 and PE 98 show a
$\bar{l} \sim 1/T$ law, which is usually observed for crystals. At 3K the mean

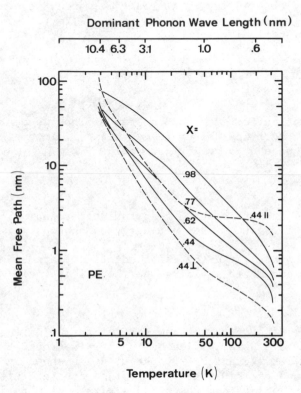

Figure 8. Mean free path $\bar{l}$ of the heat carrying phonons in semi crystalline polyethylene, calculated from thermal diffusivity data in figure 7 using eqs. (4,5). The upper scale is to indicate the dominant phonon wavelength λdom at a given temperature; λdom has been calculated from the Tarasov-model using the specific heat data of polycrystalline PE 98.

free path of all isotropic samples has reached nearly the same
value, which might be caused by phonon scattering at low temperatures
according to Little's mismatch theory[26]. As crystalline-amorphous
interfaces increase with increasing X, the influence of this
boundary scattering mechanism becomes stronger with crystallinity[27].

Drawing of a semi-crystalline polymer leads to drastic changes
in structure, which can be described by a roof-top arrangement of
lamellae with the chains aligned in draw direction[2]. Further
extrusion produces crystalline blocks pulled out of the lamellae by
tie molecules. For our drawn PE sample with a draw ratio of 6 both
roof-top lamellae and crystalline blocks do exist[21]. From the
degree of crystallinity and small angle X-ray scattering, long
period L of the isotropic and drawn PE 44 (L_{iso} = 14.5 nm; $L||$ =
10.7 nm) we conclude that the non-crystalline regions between
crystallites in draw direction vary between 1 and 2.5 nm and the
lamellae thickness is about 9.5 nm. This structure strongly
influences phonon propagation. The mean free path $\bar{l}||$ in Figure
8 has reached an exceptionally high value of 2.5 nm at 200 K, which
holds down to 50K. We believe that this behaviour is due to the
limitation of $\bar{l}||$ by the intercrystalline regions. Below 50K, the
dominant phonon wavelength λdom exceeds the noncrystalline layers,
resulting in higher $\bar{l}||$ values. Below 7K, the slope of $\bar{l}||$ (T)
starts increasing again, which might be due to λdom becoming larger
than the crystallite dimensions parallel to the extrusion direction.
Perpendicular to the extrusion direction, $\bar{l}|$ (T) is strongly
influenced by the amorphous structure. Values of $\bar{l}|$ are smaller
than that of isotropic PE 44 at high temperatures which can be understood by the alignment of chains. In the low temperature region,
the difference between $\bar{l}||$, $\bar{l}|$, and $\bar{l}$ of PE 44 decreases,
indicating that the structure changes produced by drawing become
less important.

Thermal conductivity K of isotropic PE 44, PE 62, PE 77,
PE 98 and the drawn sample has been calculated from diffusivity and
specific heat data using equation (5) (see Figure 9). In the
temperature region above 100K thermal conductivity K of the isotropic samples increases with increasing X, with the slope of K(T)
becoming negative for X >0.77. This behaviour has been explained
by the modified Maxwell model[28] which is based on a knowledge of
the conductivity of the amorphous phase K_a and the conductivity
perpendicular to the chains $K_c |$ inside the crystallites. Below
about 30K, the X-dependence of K vanishes, resulting in nearly the
same values for all isotropic samples at 3K. This is a result of
increasing boundary scattering at low temperatures. The acoustic
mismatch theory has been successfully applied to polyethylene
terephthalate by Choy and Greig[27] in order to explain increasing
thermal conductivity with increasing X at high temperatures. The
opposite trend observed below about 10K is explained by assuming
phonon scattering at the crystalline amorphous interfaces.

TABLE 2.   DEBYE–TEMPERATURES θ_D (IN K), SOUND VELOCITIES v_T, v_L (IN M/SEC) AND TRANSMISSION COEFFICIENTS Γ_T, Γ_L FOR TRANSVERSE AND LONGITUDINAL PHONONS AS USED FOR THE CALCULATION OF THE BOUNDARY RESISTANCE IN PE 98 (ACCORDING TO THE FORMULAE GIVEN BY LITTLE[26], AND CHOY AND GREIG[27]; DASHED LINE IN FIGURE 10. Γ_T, Γ_L DENOTES TRANSMISSION FROM AMORPHOUS TO CRYSTALLINE STRUCTURES (UPPER LINE) AND THE OPPOSITE DIRECTION (LOWER LINE), RESPECTIVELY.

	θ_D	v_T	v_L	Γ_T	Γ_L
*	159	1800	3600	.22	.27
$	265	2500	5000	.5	.5

* amorphous

$ crystalline

Thermal conductivities of extruded semi-crystalline polymers parallel and perpendicular to the draw direction have been discussed in detail by Choy et al[29], using the modified Maxwell model as well as the Takayanagi model (see also Greig and Hardy[18]).

To compare the thermal conductivity K of the nearly 100% crystalline sample PE 98 with that of a polymer single crystal, data for PE 98 and PTS are plotted in Figure 10 for temperatures 3 to 300K. Measurements on PTS were carried out parallel and perpendicular to the chain direction (which means parallel to the crystallographic b-direction and perpendicular to the b-c-plane). Due to difficulties in sample preparation, however, the error of these preliminary results for PTS is about 20%. To give a rough ideal of the influence of the boundary resistance R_b on K (T) of PE 98, R_b has been subtracted according to the formalism given by Little[26]; for parameters (see table 2). At about 10K the dominant phonon wavelength (see Figure 8) is expected to become equal to the inter-crystalline layers of about 4 nm and so absolute values of R_b are rather uncertain below this temperature. Due to the poly-crystalline nature of PE 98, however, acoustic mismatch between crystallites should still occur at 3K and below. Nevertheless, no principal change in K(T) is observed for PE 98 after subtracting R_b. So differences in thermal conductivity of both materials are obvious.

In contrast to PE 98, K for PTS shows only pure variation with temperature and flat maxima occur about 30K for K|| and K|. This is unexpected for a crystal and may be understood as a result of an opposite trend in temperature dependence of C_D and $\bar{l}$ according to equation (4). It is not clear at the moment, whether this lack in increase of $\bar{l}$ with decreasing temperature is due to imperfection of the PTS crystal or a result of phonon-phonon interaction in the polymer lattice. Another noteworthy feature is the anisotropy K||/K| which increases from ∿10 at 300K to ∿18 at the lowest temperature. From measurements of the elastic constants[30], an elastic anisotropy of about 6 can be estimated. Thus, at low temperatures there must be other sources contributing to the anisotropy of the thermal conductivity in PTS crystals.

THERMAL EXPANSION AND GRUNEISEN PARAMETER

As briefly discussed in the introduction of this paper, the thermal expansion of crystalline polymers has been the subject of much theoretical work, but only a few measurements down to low temperatures have been reported. One of the interesting points in studying the thermal expansion of a polymer solid is the importance of the Gruneisen-parameter which is strongly influenced by the anharmonicity of the interchain binding forces in a polymer lattice. The bulk Gruneisen parameter γb is obtained for an isotropic solid by

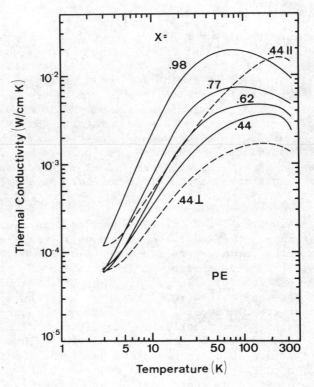

Figure 9. Thermal conductivity of semi-crystalline polyethylene, calculated from thermal diffusivity data (figure 7) and specific heat data (figure 4) using eq. (5).

$$\gamma_b = \frac{V \cdot \beta}{K_T \cdot C_v} = \frac{V \cdot \beta}{K_S \cdot C_p} \quad , \tag{6}$$

where V is the molar volume, β the coefficient of volume expansion.
K_T and K_S are isothermal and adiabatic compressibilities, and C_v
and C_p the heat capacity at constant volume and pressure. For an
anisotropic solid the number of Gruneisen parameters increases (as
the number of linear thermal expansion coefficients does). However,
in this report we discuss γ_b only, which is related to $\gamma_\perp$ (thermal
stress perpendicular to the chains) by $\gamma_b \approx \gamma_\perp$ for crystals with
axial symmetry.[3]

The temperature dependence of $\gamma_\perp$ has been studied on the anisotropic thermal expansion of chain structure crystals with axial
symmetry, i.e. trigonal tellurium and selenium[31]. Crystalline Te
and Se having a helical chain structure may be regarded as elemental
polymers although the difference between inter- and intrachain
forces is not so great as in organic polymers. The variation of $\gamma_\perp$
with temperature $(\gamma_\perp{}^{300}K = 0.85, \gamma_\perp{}^{10}K = 1.5$ for Se) is understood
on the basis of Gibbons' simple model[3] of a two-atomic linear chain
with alternating strong and weak force constants. The strong
coupling of the linear chain corresponds to the intrachain forces
whereas the weak coupling represents the interchain forces between
the polymer chains. At low temperatures only the lower frequency
vibrations, which are related to the weak coupling, are thermally
excited. Thermal stresses increase the variation of the bond
length and, hence, the corresponding frequencies depend strongly
on the volume. According to the definition of a microscopic
Gruneisen parameter for each lattice frequency

$$\gamma_i(\omega) = -\frac{\delta(\ln\omega)}{\delta v} \tag{7}$$

and

$$\gamma = \frac{\sum_i \gamma_i(\omega) \cdot C_i(\omega)}{\sum_i C_i(\omega)} \tag{8}$$

γ_b approaches relatively high positive values. At higher temperatures more and more γ_i's of high frequency vibrations (corresponding to the strong coupling with only slight variations of the bond
length) enter eq. (8). Thus a decrease of $\gamma(T)$ with increasing
temperature is observed[31]. With the linear chain model applied to
the chains in a polymer crystal Gibbons obtained a much stronger
temperature dependence of $\gamma_b(T)$. At zero K a value $\gamma_b = 4.2$ is
obtained, decreasing to $\gamma_b \approx 1$ at high temperatures. The low
temperature value is in agreement with Gruneisen parameters calculated from polymer models based on weak volume-dependent interchain
forces γ_b (0 K) = 5.25[12]; γ_b (0 K) = 3.16[14]).

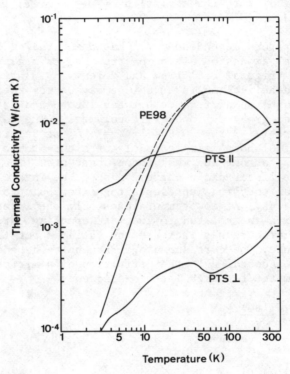

Figure 10. Thermal conductivity of polycrystalline polyethylene
and a polydiacetylene single crystal, parallel and perpendicular
to the chain direction (calculated from eq. (5)). Below 50K
the thermal resistance due to internal boundary scattering
(Choy and Greig[27]) has been substracted (dashed line).

Another point of view has been given by Wada et al[13], by describing thermal properties related to the anharmonic potential in terms of interchain vibrations. By measuring the pressure dependence of the sound velocity around 300K Wada et al. obtained γ values of about 3 to 4 on PE, PMMA and PS. These high values occur because sound wave propagation excites only interchain modes in the polymer and the appropriate heat capacity is the interchain contribution only. The same argument is still valid at low temperatures where only interchain vibrations are contributing to the specific heat. Thus, one can expect from Wada's model that Gruneisen parameters calculated according to eq. (6) from low Temperature C_p and β-data of crystalline polymers reveal high values.

In Figure 11 the linear expansion coefficients and the related Gruneisen functions γ_b for semi-crystalline PE are shown in the temperature range 10K to 300K. So far, precise α -data down to 10K is available only for two samples, PE 59 and PE 77, using our inter-ferometric dilatometer[31]. Above 100K a thermomechanical method was used in addition, yielding α-data for all PE samples (X = 0.44 to 0.98) and, recently, for the three crystallographic directions of the PTS single crystal (not shown in Figure 11). In order to complete the low temperature data of PE we have used α-values of low density PE, reported by Barker[32], which agree quite well with our measurements on PE 44 between 100 and 300K.

In order to calculate the Gruneisen parameter, eq. (6), the bulk compressibilities K_T of PE and PTS are necessary. For PE the K_T values depend on crystallinity, as has been shown by Hellwege et al[33]. Because there is a lack of temperature dependent K_T data we used the data of Hellwege et al at 300K assuming that K_T is not affected by the temperature significantly. For PTS we used $K_T = 5.3 \cdot 10^{-11}$ m^2/N, calculated by Bassler[34] from the variation of the PTS unit cell under pressure at 300K.

Comparing γ_b of PTS and PE 98 (Figure 11) both Gruneisen functions increase with decreasing temperature. At 100K a value $\gamma_b \simeq 3$ for PTS and $\gamma_b \simeq 1.5$ for PE 98 has been calculated. The bulk Gruneisen parameter for PTS seems to be in quite good agreement with the current theories, however, the low temperature dependence and the zero K value may be of great interest. For PE 98 the value of γ_b seems to be too low and the situation is somewhat unclear as has been stated by White et al [9] for a POM crystal. Because the Gruneisen functions of PE 98 and POM are in good agreement in the temperature range 100 to 300K, we also assume, that there could be cross linking between the chains and thermal stresses between the crystallites to influence significantly the thermomechanical behaviour of PE 98. In view of the low temperature data of semi-crystalline PE it would be of great interest to measure the thermal expansion of PE 98 down to low temperatures.

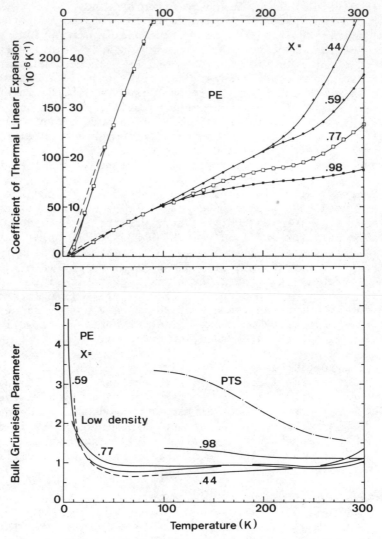

Figure 11. Coefficient of thermal expansion α and Gruneisen
function γ_b of semi-crystalline polyethylene. Below 100K
measurements of α have been performed only by an interferometer
method, above 100K a thermomechanical method was used in
addition. For low density PE the low temperature data of
Barker[32] are shown (dashed line). The Gruneisen functions have
been calculated using eq. (6); the bulk compressibilities K_T
at 300K have been taken from Hellwege et al[33] and were
assumed to be temperature independent. The bulk Gruneisen
function of a PTS single crystal has been calculated from
recent α-measurements in three crystallographic directions
(Ershad[39]) and K_T data given by Bassler[34].

The thermal expansion of semi-crystalline PE above 120K and below 50K is strongly affected by its amorphous phase. In the high temperature region α increases with temperature and amorphous fraction, which is ascribed to trans-gauche conformational changes and, above 240K to hole contributions[1]. Below 50K the thermal expansion of PE decreases with crystallinity as can be seen from the data of PE 77, PE 59, and low density PE. The "excess" in thermal expansion due to the amorphous matrix is of the same order as has been observed for C_p low temperature data (Figure 5). Calculating the realted Grüneisen functions a similar behaviour for all three γ_b's is observed: at 50K γ_b 0.9 ± 0.1, increasing to γ_b 2.2 ± 0.2. For PE 59 a Gruneisen parameter γ_b 4.5 ± 1 at T = 5K has been calculated.

The strong increase of γ_b of semi-crystalline PE below 20K may support the current theories yielding γ 4 ± 1 at 0 K. On the other hand, assuming that the low frequency modes - characteristic to the amorphous state and responsible for the excess specific heat around 5K- are of strong anharmonic character, the resulting microscopic Gruneisen parameters may also give rise to the observed increase of γ_b. Thus, the problem is not specific to polymer chains, but more related to anharmonic excitations of the amorphous state. To clear up the situation and for any detailed understanding of the anharmonic interactions in crystalline and amorphous polymers, more low temperature measurements of the thermal expansion of crystalline polymers are essential.

ACKNOWLEDGEMENTS

The authors wish to thank Dr M Pietralla (Ulm) and Dr H Bässler (Marburg) for many helpful discussions and for supplying the PE 98 sample and several PTS crystals. The authors are also grateful to Dr H Baur (Ludwigshafen) and Dr D Greig (Leeds) for giving valuable comments and suggestions. All other PE samples were kindly made available to us by BASF AG (Ludwigshafen).

REFERENCES

1.   Wunderlich B and Baur H: Adv. Polym. Sci. 7, 151 (1970.

2.   Choy C L: Polymer 18, 984 (1977) and references cited in.

3.   Gibbons T G: J. Chem. Phys. 60, 1094 (1974).

4.   Pohl R O and Salinger G L: Ann. N.Y. Acad. Sci. 279, 150 (1976).

5.   Phillips W A: J. Low Temp. Phys. 7, 351 (1972);
      Anderson P W, Halperin B I and Varma C M: Phil. Mag. 25, 1 (1972)

6.   Federle G and Hunklinger S: this proceedings (1980).

7. Schickfuss M and Hunklinger S: this proceedings (1980).

8. Chang S.S: J. Res. Nat. Bur. Stand. (U.S.) 78A, 387 (1974).

9. White G K, Smith T F and Birch J A: J. Chem. Phys. 65, No.2, 554 (1976).

10. Baur H: Kolloid Z. Z. Polym. 244, 293 (1971); 241, 1057 (1970).

11. Barker R E: J. Appl. Phys. 38, 4234 (1967).

12. Broadhurst M G and Mopsik F I: J. Chem. Phys. 52, 3634 (1970).

13. Wada Y, Itani A,Nishi T and Nagai S: J. Polymer Sci. A-2, 7, 201 (1969).

14. Curro, J G: J. Chem. Phys. 58, 374 (1973).

15. Choy C L, Hunt R G and Salinger G L: J. Chem. Phys. 52, 3629 (1970).

16. Engeln I and Meissner M: J. Polymer Sci. (Polym. Phy. Ed.), in the press (1980).

17. Baur H: Kolloid Z.Z. Polym. 250, 1000 (1972).

18. Greig D and Hardy N D: this proceedings (1980).

19. Schermann W, Wegner G, Williams J O and Thomas J M: J. Polym. Sci. (Polym. Phys. Ed.) 13, 753 (1975).

20. Stockmayer W H and Hecht C E: J. Chem. Phys. 21, 1954 (1953).

21. Kanig G: Prog. Colloid and Polymer Sci. 57, 176 (1975).

22. Tucker J E and Reese W: J. Chem. Phys. 46, 1388 (1967).

23. Meissner M and Wobig D: Proc. Int. Conf. on the Physics of Se and Te (ed. E Gerlach, P Grosse), Springer (Berlin), p 68 (1979).

24. Zirke J and Meissner M: Infrared Physics 18, 871 (1978).

25. Kruger R, Meissner M, Mimkes J and Tausend A: Phys. stat. sol. (a) 17, 471 (1973).

26. Little W A: Can. J. Phys. 37, 334 (1959).

27. Choy C L and Greig D: J. Phys. C8, 3121, (1975).

28. Choy C L and Young K: Polymer 18, 769 (1977).

29. Choy C L, Luk W H and Chen F C: Polymer 19, 155 (1978).

30. Leyrer, R J, Wegner G and Wettling W: Ber. Bunsenges. Phys.
     Chem. 82, 697 (1978).

31. Grosse R, Krause P, Meissner M and Tausend A: J. Phys. C11, 45
     (1978) and references cited in.

32. Barker R E: J. Appl. Phys. 34, 107 (1963).

33. Hellwege K H, Knappe W and Lehmann P: Kolloid. Z.Z. Polym. 183,
     110 (1962).

34. Bassler H, unpublished results (1980).

35. Zeller R C and Pohl R O: Phys. Rev. B4, 2029 (1971).

36. Lasjaunias J C, Maynard R and Thoulouze D: Sol. State Com. 10,
     215 (1972).

37. Stephens R B, Cieloszyk G S and Salinger G L: Phys. Let. 38A,
     215 (1972).

38. Hager N E jr: Rev. Sci. Instr. 31, 177 (1960).

39. Ershad A: Diplomarbeit, TU Berlin (1980).

THERMAL CONDUCTIVITY AT LOW TEMPERATURES

IN SEMICRYSTALLINE POLYMERS

D Greig and N D Hardy

Department of Physics
University of Leeds
Leeds LS2 9JT, UK

Over the past few years we have studied in detail the influence of (i) crystallinity and (ii) crystallite orientation in determining the thermal conductivity, κ, of semicrystalline polymers.[1-3] Between about 2K and room temperature the variations in κ with these two parameters are completely different above and below about 25K. At the higher end of the range the conductivity increases both with crystallinity and with orientation, the latter being brought about by drawing or extruding the isotropic polymer. At low temperatures, on the other hand, the specimens with the greatest crystallinity have the lowest conductivity with values in a 50% crystalline specimen at $\sim$ 1.5K roughly an order of magnitude lower than those found "universally" for amorphous solids.[4] These low values, furthermore, are found to be more or less independent of crystallite orientation. The overall behaviour just described is shown schematically in figure 1.

This rather dramatic change in behaviour is attributed to the different conditions arising at low temperatures when the phonon mean free path, ℓ, becomes larger than the dimensions, L, of the intercrystalline (amorphous) structural units. As the difference in density between crystalline and amorphous regions can be as great as 20% the "structure scattering" of the composite polymer gives rise to a thermal resistance that is considerably greater than that of the amorphous material alone.

At still lower temperatures, somewhere in the region of 1K, it is found that, for polyethylene the variation of κ with T undergoes a rather sharp decrease in slope, changing from a dependence in the temperature range 1K to 20K of about $T^{1.8}$ to a linear dependence below that temperature.[3,5,6] This feature has been explained on

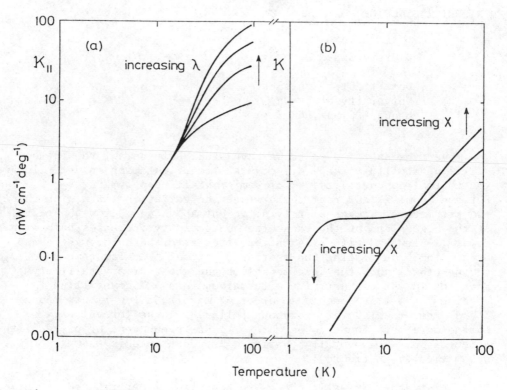

Figure 1. Schematic temperature dependence of thermal conductivity (a) parallel to the extrusion direction for various
degrees of extrusion, λ and (b) with crystallinity, X.

the basis of the changing conditions when the dominant phonon wave-length, $\bar{\lambda}$, becomes greater than the thickness of the crystallites, and has been found to occur at rather higher temperatures in a specimen of polyethylene that has been extruded. This particular feature of the results has not, however, been studied in detail; for example, we have not, as yet examined the variation of κ below 1K with crystallinity or with extrusion other than at extrusion ratios that are very low.

In this short paper we shall give some details of the models that have been proposed to explain the results in the three temperature ranges outlined above; namely above 25K, between 2K and 25K, and at very low temperatures. In addition we shall present one or two new results to give some appreciation of behaviour that is "universal" and trends that only apply to a particular polymer.

TEMPERATURES ABOVE 25K

At these temperatures κ increases with crystallinity and is very dependent on the relative orientation of the crystallites. In figure 2 we show the values of conductivity at 100K in directions both parallel and perpendicular to the extrusion directions in Rigidex 50 – – a linear polyethylene with $\bar{M}_w = 1.01 \times 10^5$. It is clear that whereas the conductivity parallel to the extrusion direction, $\kappa_\parallel$, increases slowly with extrusion ratio λ, the perpendicular conductivity, $\kappa_\perp$, is about half the isotropic value and more or less independent of λ. The value of κ in every case is clearly governed by the role of the crystallites and this arises because the phonon mean free path ℓ and dominant wavelength, $\bar{\lambda}$, are both much smaller than the intercrystallite dimension, L. In this temperature region the overall conductivity can therefore be explained by a "classical" model of crystalline and amorphous regions in series and parallel with the conductivity in the crystalline regions very highly anisotropic. (Our estimates have shown $\kappa_\parallel / \kappa_\perp$ in the crystallites at 100K to be somewhere in the region of 50.)[7]

In two earlier papers[2,8] we have discussed a remarkable one-to-one experimental correlation between the variation of $\kappa_\parallel$ with λ at 100K and the increase with extrusion of Young's modulus, $E_\parallel$, at temperatures between the α and γ relaxations. We deduced that the compliance and the thermal resistance of the material are both ultimately determined by the intercrystalline regions and that the continuing increase in both properties with λ arise (a) at low extrusion ratios from crystallite orientation and (b) at high extrusion ratios by the creation of intercrystalline bridges. In the latter case we have used a modified Takayanagi model to explain the behaviour by what is essentially a "geometrical" argument.[8] As shown in figure 3 the material may be regarded as an inter-crystalline bridge fraction (1-f) which is continuous throughout

the specimen forming a parallel path with the remainder of the
material which is represented by a lamellar fraction $(1-\phi)$ in series
with a non-crystalline fraction ϕ. If we assume that the bridges
are crystalline in nature with the same high conductivity in the
extruded direction as the crystals themselves, then it is easy to
show that,

$$\kappa_{||}/\kappa_{c_{||}} = E_{||}/E_{c_{||}} \simeq (1-f) \qquad\qquad (1)$$

thus showing the proportionality between $\kappa_{||}$, $E_{||}$ and the fraction
of intercrystalline bridges.

TEMPERATURES BETWEEN 2K and 25K

At these lower temperatures where $\ell > L$ the situation is completely reversed. Polymers that are about 50% crystalline have by
far the lowest conductivity with values that are independent of the
orientation of the crystallites or the presence of intercrystalline
bridges. The main role of the crystallites is now to provide
regions of higher density from which phonon are reflected by acoustic
mismatch.

To discuss this problem in detail, two models have been used.
Choy and Greig[1] have discussed the problem in terms of a model first
developed by Little[9] in which the thermal boundary resistance, r_b,
between the crystalline and amorphous regions is expressed in terms
of the ratios of their densities and sound velocities. For crystallites of a fixed thickness d and cross sectional area the number
of boundaries per unit length is 2X/d; that is the number is
proportional to the crystallinity, X. Consequently the total
boundary resistance is given by $2Xr_b/d$, so that the overall conductivity is roughly inversely proportional to X. Choy and Greig[1]
found that, for a series of specimens of PET for which X varied
from 0 to 0.51 (51%) the measurements at 2K fitted this model
extremely well. As this boundary resistance varies as T^{-3} it is
clear that at higher temperatures this effect becomes negligible
and the situation reverts to that discussed in the previous section.

The disadvantage of this model is that it completely fails to
account for the changes in slope observed in the κ versus T curves
at still lower temperatures. For this the "structure scattering"
model as developed by Klemens,[10] Ziman,[11] Morgan and Smith,[12] and
Walton[13] for diffraction of lattice waves by regions of different
density seems more appropriate. Assfalg[14] has analysed the variation of κ with X in PET by assuming that the structure factor for
velocity fluctuations is proportional to the structure factor for
electron density fluctuation which he obtained by low-angle X-ray
diffraction, and has obtained good agreement with the measured
data.

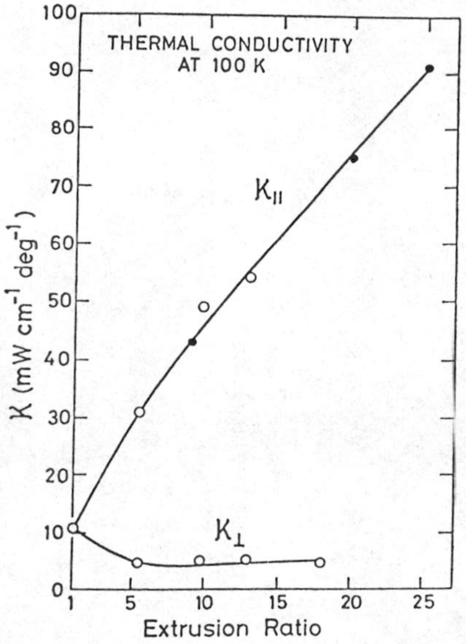

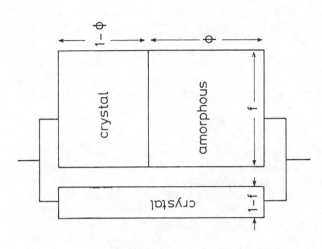

Figure 3. Modified Takayanagi model with the intercrystalline bridge fractions (1-f) in parallel with the rest of the specimen. The fraction φ represents the amorphous material between lamellae.

Figure 2. Variation of thermal conductivity at 100K for various values of extrusion ratio, λ : O, small diameter extrudates; O specimens cut from large diameter extrudates.

A detailed calculation by Turk and Klemens[15] has shown that
for 'plates' of thickness d the phonon mean free path in conditions
where $\bar{\lambda} \ll d$ is given by

$$\ell \simeq d/XR \qquad\qquad (2)$$

where R is the phonon reflection coefficient arising from the
acoustic mismatch referred to above. Since, from simple kinetic
theory, $\kappa \alpha \ell$, we see that once again κ is inversely proportional to X.

TEMPERATURES BELOW 2K

At really low temperatures the amount of data on semicrystalline polymers is very sparse. Nevertheless, as mentioned above,
3 independent measurements on different specimens of polyethylene
have shown an abrupt reduction in slope of κ versus T as the
temperature is lowered below about 0.5K. For an extruded rod of
rather low extrusion ratio ($\lambda = 3.9$) the change in slope occurs at
about 2K. The various authors agree that this anomaly occurs when
$\bar{\lambda}$ becomes greater than the dimensions of the crystallites. The
diffraction conditions are then changed and become rather analogous
to Rayleigh scattering from point scatterers. For plates embedded
in an isotropic amorphous matrix Turk and Klemens have shown[15]

$$\ell = 3v^2 \{8\pi d\omega^2 (\Delta v/v)^2 X\}^{-1} \qquad\qquad (3)$$

where ω is the angular frequency of the phonons of velocity v. The
factor Δv is the difference of velocity in the amorphous and
crystalline regions. Bhattacharyya and Anderson[6] have shown that
the change of slope can be easily explained by this new form of ℓ
at very low temperatures and that the experimental effect persists
to as low as 30 mK.

NEW RESULTS; OTHER SYSTEMS

In an earlier publication[16] we showed that, at temperatures
between 1.5K and 100K, the ideas developed in the previous sections
based on measurements on polyethylene and polyethylene terephthalate
could be applied equally well to extruded polypropylene and extruded
polyoxymethylene. There are, of course, differences in detail but
the qualitative trends shown in figures 1a and 1b are still
maintained. However, so far as we are aware, no measurements have
been made below 1K on any semicrystalline polymer other than polyethylene.

To conclude the paper we should like to present some new data
in the temperature range 2.5K to 100K on two samples of Rigidex 40
- - a copolymer containing 5 methyl side groups per thousand main
chain carbon atoms. As Hope et al[17] had found the values of $E_{\parallel}$
in this system to be considerably smaller for a given extrusion

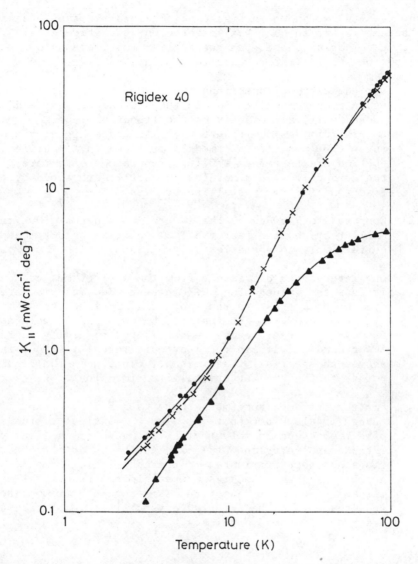

Figure 4. Temperature dependence of thermal conductivity of the copolymer, Rigidex 40. Δ , isotropic specimen; +, extrusion ratio 17; 0, extrusion ratio 20.

ratio than those in the samples of Rigidex 50 which have been studied
in such detail, it was clearly of great interest to find if the
correspondence with $\kappa_{||}$ still existed.

The results are shown in figure 4. It is clear that once again
the measurements are in qualitative agreement with figure 1a.
However, in comparing these with the earlier measurements on
Rigidex 50[2] we should like to make four points:

(1) For the extruded specimens the values of $\kappa_{||}$ at 100K are
     again proportional to $E_{||}$ in the plateau region, and
     indeed still fit exactly on the linear plot of $\kappa_{||}$ versus
     $E_{||}$ for Rigidex 50 (figure 4[2]). At these higher tempera-
     tures the 'geometrical' models outlined in an earlier
     section are therefore still appropriate and we infer that
     the conductivity and modulus are still determined by the
     fraction of intercrystalline bridges.

(2) Compared to Rigidex 50 the isotropic values at 100K are
     smaller by about a factor of 2. This can be attributed
     quite simply to a smaller crystallinity in the copolymer.

(3) One interesting difference from the earlier measurements
     is that the values of $\kappa_{||}$ in the extended material are
     greater than those in the isotropic specimen throughout
     the temperature range; that is, below $\sim$ 25K as well as
     above. This feature may be attributed partially to the
     lower conductivity in the isotropic material as outlined
     above, and partially, as we shall explain in the final
     point, to the smaller crystallite dimensions.

(4) For the extruded specimens there is a fairly abrupt
     change in slope at about 7K, almost identical to the
     effect observed in polyethylene at 0.5K and in lightly
     extruded polyethylene at 1.5K. As these earlier observa-
     tions had been attributed to the changing diffraction
     conditions when $\bar{\lambda}$ becomes comparable to the crystallite
     thickness, we shall adopt the same interpretation here
     and deduce that the oriented crystallites in Rigidex 40
     are probably 40-50 Å thick.

SUMMARY

The temperature dependence of κ in semicrystalline polymers in
the temperature range 1.5K to 100K is well represented by figures 1a
and 1b. At lower temperatures there are abrupt changes in the slope
of κ versus T that have still to be investigated in detail.

From the point of view of the structure and physics of polymers
the detailed variation in the curves from one series to the next

tells us something about the crystallinity of the material as well as about the crystallite dimensions and the presence of intercrystalline bridges in the specimen. As regards low temperature applications there is not a huge difference in the values of κ between the various systems although at 3K for example, by a careful choice of material it is possible to select conductivities that are a factor of 2 greater or smaller than the 'norm' of 0.1 mW cm^{-1} deg^{-1}.

ACKNOWLEDGEMENTS

We are greatly indebted to Professor I M Ward and his colleagues at Leeds for the continuing supply of well characterised polymers and to the Science Research Council for considerable financial support.

REFERENCES

1.   C L Choy and D Greig, J. Phys. C., Solid State Physics, 8, 3121-30 (1975).
2.   A G Gibson, D Greig, M Sahota, I M Ward and C L Choy, J. Polym. Sci., Polym. Letts. Ed., 15, 183-92 (1977).
3.   D M Finlayson, P Mason, J N Rogers and D Greig, J. Phys. C., Solid State Physics, 13, L185-8 (1980).
4.   R B Stephens, Phys. Rev. B8, 2896-905 (1973).
5.   M Giles and C Terry, Phys. Rev. Lett., 22, 822-3 (1969).
6.   A Bhattacharyya and A C Anderson, J. Low Temp. Phys., 35, 641-6 (1979).
7.   S Burgess and D Greig, J. Phys. C., Solid State Physics, 8, 1637-48 (1975).
8.   A G Gibson, D Greig and I M Ward, J. Polym. Sci., Polym. Phys. Ed., in the press (1980).
9.   W A Little, Can. J. Phys., 37, 334-49 (1959).
10.   P G Klemens, Proc. R. Soc., A208, 108-33 (1951).
11.   J M Ziman, Electrons and Phonons (Oxford: Clarendon Press) (1960).
12.   G J Morgan and D Smith, J. Phys. C., Solid State Physics, 7, 649-64 (1974).
13.   D Walton, Solid St. Commun. 14, 335-9 (1974).
14.   A Assfalg, J. Phys. Chem. Solids, 36, 1389-96 (1975).
15.   L A Turk and P G Klemens, Phys. Rev. B9, 4422-8 (1974).
16.   C L Choy and D Greig, J. Phys. C., Solid State Physics, 10, 169-79 (1977).
17.   P S Hope, A G Gibson and I M Ward, J. Polym. Sci., Polym. Phys. Ed., in the press (1980).

ELASTIC NONLINEARITY OF VITREOUS SILICA

AT LOW TEMPERATURES

M v Schickfus, S Hunklinger and K Dransfeld

Max-Planck-Institut für Festkörperforschung
Heisenbergstr. 1
D-7000 Stuttgart 80, Federal Republic of Germany

INTRODUCTION

Low-temperature anomalies in the thermal, acoustic and die-lectric properties of amorphous solids have found increasing atten-tion in recent years.[1] Compared with crystals, these materials exhibit an anomalously high specific heat and low thermal conduc-tivity.[2] This behaviour can be explained by the existence of low-energy excitations with a nearly constant density of states in an energy range extending at least between 10 mK and 10 K. More detailed information has been obtained by ultrasonic and dielectric measurements[3] which demonstrate that these excitations are best described as two-level systems. According to a generally accepted model they originate from small groups of atoms, which are ener-getically nearly equivalent, tunnelling between two positions in the amorphous matrix.[4] About the physical nature of the tunnelling "particles", however, not much has previously been known.

Since two-level systems exhibit extremely anharmonic properties, we expect strong elastic nonlinearities in amorphous solids at low temperatures, where these excitations determine their thermal properties. In order to investigate these nonlinearities, we have carried out the following experiment: At low temperatures (T < 15 K) a sample of vitreous silica is strained mechanically. Because of the nonlinearity of the amorphous substance this will lead to a temperature variation via the thermoelastic effect. From the measured temperature change we are able to determine the Grüneisen parameter. In addition an irreversible heating effect is observed from which we are able to obtain information on the creep behaviour of glasses at these low temperatures.

THEORETICAL

Generally the nonlinear properties of solids can be expressed by the Grüneisen parameter γ. For an excitation of energy E_i the corresponding Grüneisen parameter is given by:

$$\gamma_i = - \frac{\partial \ln E_i}{\partial \ln \rho} = - \frac{\delta E}{E} \cdot \frac{1}{e} \tag{1}$$

where ρ is the mass density and e the strain (for simplicity we only consider pure dilation here). This microscopic Grüneisen parameter reflects the pressure dependence of the energy of the excitations under consideration. A value for γ derived from macroscopically observable quantities can be expressed in terms of thermal expansion β, compressibility χ, and specific heat c_v (per volume):

$$\gamma = \frac{\beta}{\chi \cdot c_v} \tag{2}$$

Since γ as well as β are roughly constant for most solids, the thermal expansion of dielectric crystals decreases at low temperatures with the cube of temperature as their specific heat. Since furthermore γ is of the order of unity for acoustic phonons, thermal expansion becomes exceedingly small and, to our knowledge, no measurements below 1.5 K have been performed so far. In a glass, however, the situation is different since tunnelling states give rise to additional anharmonic modes.

In order to reduce the number of parameters in our theoretical description, we assume that the tunnelling system can be described as a "particle" moving in a potential consisting of two adjacent parabolic wells (see Figure 1). In this case there is a simple relation between the barrier height V, the distance 2d between the wells, the mass m, and the oscillation frequency $\Omega/2\pi$ of the particle in one well:

$$\Omega^2 = 2V/md^2 \tag{3}$$

The energy splitting E of the tunnelling system is composed of two contributions, namely the asymmetry Δ in the depth of the wells and the tunnel splitting Δ_o due to the overlap of the wavefunction of the particle in the two wells:

$$E^2 = \Delta_o^2 + \Delta^2 \tag{4}$$

The tunnel splitting Δ_o can also be expressed by the parameters introduced above:

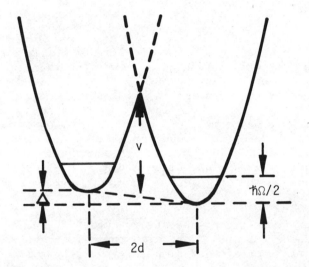

Fig 1. The double-well potential of the tunnelling model.

$$\Delta_o = n\Omega \sqrt{\frac{\lambda}{\pi}} \, e^{-\lambda} \tag{5}$$

where $\lambda = (8md^2V/n^2)^{\frac{1}{2}}$. Because of the randomness of the amorphous structure the value of Δ and λ will not be well defined, but will have a broad distribution. For small values of E it is generally assumed that Δ as well as λ exhibit a constant density of states, ie $P(\Delta,\lambda)d\Delta d\lambda = \bar{P}d\Delta d\lambda$ where $\bar{P}$ is a constant. This distribution can only be valid within a certain range of λ. The lower limit λ_{min} is given by the condition $\Delta_o \leq E$. The upper limit λ_{max} is less obvious but has to be introduced in order to be able to normalise this density of states. According to equations 1 and 4, the strain dependence of E is composed of the strain dependence of Δ and of Δ_o, resulting in two contributions to the Grüneisen parameter:

$$\gamma = -\frac{1}{e} \left(\frac{\Delta_o \delta\Delta_o + \Delta\delta\Delta}{E^2} \right) \tag{6}$$

The strain can vary the asymmetry Δ in two different ways. Firstly, it can directly influence Δ by varying the environment of the tunnelling system. This process is probably the dominant coupling mechanism in ultrasonic experiments. For this change in the asymmetry we can write: $\delta\Delta = B_\Delta \cdot e$. The deformation potential B_Δ is known to be very large, namely of the order of 1 eV resulting in a value of $\gamma_{\Delta 1}$ of the order of 10000 for level splittings around $E/k = 1$ K. Macroscopically, however, this mechanism does not contribute to the thermal expansion, since Δ and B_Δ are not correlated and the contribution of the individual tunnelling systems cancel each other.

The second influence of the strain onto Δ is due to the fact that it varies the distance 2d between the wells. We assume that Δ and d are correlated and for simplicity we take a linear relationship:

$$\Delta = K \cdot d \tag{7}$$

where K is a constant. With $\delta d/d = -e/3$ we then obtain

$$\delta\Delta = K \cdot \delta d = -\Delta \cdot e/3 \tag{8}$$

and

$$\gamma_{\Delta 2} = \frac{1}{3} \left(\frac{\Delta}{E} \right)^2 \tag{9}$$

Since $\Delta \leq E$ by definition, the upper limit for $\gamma_{\Delta 2}$ is 1/3.

Now we turn to the first term in equation 6 describing the contribution of Δ_o to the Grüneisen parameter. Following Lyon et al[6] we find:

$$\gamma_{\Delta o} = \gamma_{\Omega} \left(\frac{3}{2} - \lambda\right) + \frac{2}{3}\lambda - \frac{1}{3} \tag{10}$$

The effect of the applied strain is twofold; Firstly, it varies the distance d and thus the barrier height V. Secondly, it changes the shape of the potential wells. The latter effect is taken into account by introducing γ_{Ω} which is expected to be "normal", ie $\gamma_{\Omega} \simeq 2$. Since γ should vary[7] between 4 and 12, we expect $\gamma_{\Delta o}$ to be about −20.

The macroscopic value of γ is calculated by weighting the contribution of all tunnelling states with their corresponding specific heat:

$$\gamma = \frac{\Sigma\gamma_i c_i}{\Sigma c_i} \qquad \frac{\int\int P(\Delta,\lambda)\ \gamma(\Delta,\lambda)\ c(\Delta,\lambda)\ d\Delta d\lambda}{\int\int P(\Delta,\lambda)\ c(\Delta,\lambda)\ d\Delta d\lambda} \tag{11}$$

where $c(\Delta,\lambda)$ is the contribution of the tunnelling systems with the parameters Δ and λ to the specific heat. For the constant distribution $\bar{P}d\Delta d\lambda$ introduced above this equation can be approximated by

$$\gamma \simeq 2.23 \frac{\gamma_{\Delta o}}{\ln\left(\dfrac{6kT}{\Delta_o^{min}}\right)} \tag{12}$$

where Δ_o^{min} is the tunnel splitting corresponding to λ_{max}. This expression is only valid in the limit of very low temperatures (T < 0.3 K) where the specific heat due to the tunnelling states is really linear and the contribution of the phonons is negligible. At higher temperatures we have to take into account that the specific heat due to the tunnelling states increases more rapidly than linear[1] and that furthermore the contribution of the phonons becomes important. The Grüneisen parameter, therefore, is expected to increase rapidly with temperature. The exact temperature dependence is, however, not predictable, since only the total specific heat is known and not the contribution from the tunnelling states or phonons separately.

As mentioned above, it is difficult to measure the thermal expansion at low temperatures, and therefore we used a different approach: Instead of measuring the change in length of the sample with temperature, we applied an external stress σ and measured the associated variation of temperature. The correspondence between γ, the applied stress and the temperature variation is derived as follows. For a reversible process the change in entropy S is given by:

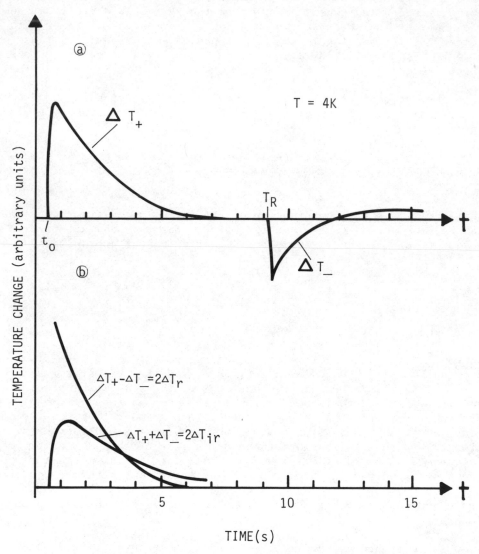

Fig 2. a) Temperature variation after pulling and releasing the
           stress.
        b) Temperature variation due to the reversible (ΔT_r) and
           the irreversible (ΔT_{ir}) thermoelastic processes.

$$dS = (\frac{\partial S}{\partial T})_V \, dT + (\frac{\partial S}{\partial V})_T \, dV = c_v \frac{dT}{T} + \frac{\beta}{\chi} \frac{dV}{V} \tag{13}$$

If furthermore the process is performed adiabatically, the entropy stays constant, ie

$$c_v \frac{dT}{T} = - \frac{\beta}{\chi} \frac{dV}{V} \tag{14}$$

Applying equation 2 for a stress which is applied unidirectionally, we obtain:

$$\gamma = - \frac{3}{T \cdot \chi} \cdot \frac{\Delta T}{\Delta \sigma} \tag{15}$$

EXPERIMENTAL

Our sample of vitreous silica (Suprasil I, length 30 cm, diameter 0.3 cm) was mounted in the vacuum chamber of a ^{4}He cryostat through a stainless steel wire with a pneumatic system. All seals were made with bellows to reduce friction, and the pulling force (10 to 100 N) was measured inside the cryostat using a strain gauge. With this system stress could be applied within 50 msec. Three thermometers were mounted on the sample: One in the centre to measure the thermoelastic effect and two at the ends to monitor possible heating effects due to friction in the mountings (this was found to be negligible). We used 220 Ω Allen-Bradley Resistors which were ground down to a diameter of 1 mm to reduce thermal mass and to improve thermal contact. With a simple Wheatstone bridge and a PAR 124 Lock-In amplifier as a detector this system reached a temperature resolution of 10^{-5} K and a typical response time of 0.7 sec at 4 K.

The output of the Lock-In was applied to a signal averaging system which recorded the temperature variation after applying and releasing the stress. Figure 2a shows a typical pattern obtained at 4 K: Temperature rises on tensioning the sample and drops after releasing the stress, corresponding to a negative Grüneisen parameter. It is clearly recognisable, however, that the temperature rise after applying the stress is larger than the temperature drop after releasing it. This is in contrast to the prediction of equation 15 from which we expect a reversible thermoelastic effect. Consequently, a second effect is observed, which is not reversible. It produces heat on both tensioning and releasing.

In order to separate the two contributions, the portion of the curve after t_R was shifted to t_0 and their sum and difference were formed (Figure 2b). The difference $\Delta T_+ - \Delta T_- = 2\Delta T_r$ represents the contribution of the reversible thermoelastic process, and the sum $\Delta T_+ + \Delta T_- = 2\Delta T_{ir}$ represents the irreversible contribution.

One could argue that the observed irreversible heating effect was due to friction in the mountings. We have checked this question very carefully by varying the length of the sample, using different types of mountings, measuring the temperature close to the mountings (where the temperature variation turned out to be smaller than at the centre of the sample due to thermal anchoring), trying to simulate the irreversible effect by electrical heating of the mountings and by varying the thermal relaxation times by letting in exchange gas. The result of all these studies is that the irreversible heating effect is in fact due to the sample itself. By measuring at 2 K where this effect dominates, we were able to show that the temperature rise due to pulling is exactly the same as in the case of releasing.

DISCUSSION

Let us first consider the reversible thermoelastic process: The temperature rise due to this process is so rapid that our system could not resolve it. At 4 K, because of thermal conduction of the sample, equilibrium temperature was restored within 30 sec after an exponential decay. The thermoelastic temperature change was determined by extrapolating to $t = t_0$. Figure 3 shows the temperature dependence of the Grüneisen parameter derived from these data. Within 10% our data agree with those published by other authors.[6,8] As expected the low temperature limit of the Grüneisen parameter given by equation 12 is not yet reached in our experiments. Nevertheless we observe a steady decrease with temperature in agreement with the expectations from the tunnelling model.

Let us now turn to the irreversible heating effect. From Figure 2b it is obvious that in this process heat is generated at a much slower rate, so that the first part of the curve may also be resolved. A slower time constant appears in the decaying part as well indicating that heat is still generated after some 10 seconds. An analysis of the temperature profile shows that independent of temperature about 5×10^{-3} erg/g are generated at a stress of 1 bar on tensioning as well as on releasing.

This result can be interpreted in terms of tunnelling systems with very long relaxation times. The relaxation time τ of a tunnelling system is given [3,9] by:

$$\tau^{-1} = A \cdot \Delta_o^2 \cdot E \coth(E/2kT) \tag{16}$$

Here A is a constant reflecting the coupling to the amorphous network. Since λ and therefore Δ_o exhibit a distribution of their values, we will also find a distribution of relaxation times even for tunnelling systems of a fixed energy splitting. For a simplified description we divide the tunnelling systems into three groups. For the first group the condition $\tau < t_s$ holds, where t_s is the

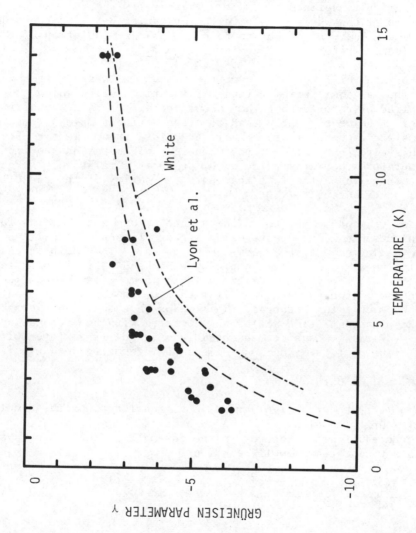

Fig. 3. Temperature dependence of the Grüneisen parameter. The results of White[8] and Lyon et al.[6] are included for comparison.

time during which the strain is raised or released – roughly 50 msec
in our experiment. Tunnelling systems belonging to that group will
always be in thermal equilibrium. They alone contribute to the
reversible part of the thermoelastic effect for which equations 14
and 15 hold. As mentioned above, the magnitude of this effect is
proportional to $\gamma_{\Delta o}$ since the contribution of B_Δ cancels. The
second group is defined by $\tau > t_a$, where t_a is the thermal relaxation time of the whole sample in the cryostat, being of the order
of 10 sec. Their contribution to a temperature variation of the
sample can therefore not be detected.

For the third group we have $t_s < \tau < t_a$. These systems do not
reach thermal equilibrium during t_s, the time during which the stress
is applied or removed, but they slowly relax into a new equilibrium
during t_a and consequently lead to creep of the sample. These
systems are responsible for the irreversible thermoelastic effect.
For this effect the sign of the Grüneisen parameter has no importance
and therefore the large value of B_Δ produces a relatively strong
effect in spite of the fact that only a minority of the tunnelling
states contributes.

Now we want to put these ideas into a more quantitative form.
From ultrasonic measurements we know that glasses behave like
viscoelastic media even at low temperatures.[9] In a simplified
description we replace our sample by a Hookean solid in parallel
with a Maxwell body.[10] The stress in the sample can be considered
to consist of two parts: The first part σ_1 is the stress brought
about by slow volume changes. This means that these changes have
to take place on a time scale long compared with the relaxation
time τ. First we consider an ensemble of tunnelling states with a
fixed value of τ. For our one-dimensional geometry we can write:

$$\sigma_1 = Y_1 \cdot \delta\ell/\ell \tag{17}$$

where $\delta\ell/\ell$ is the relative variation of the length of the sample
and Y_1 Young's modulus. The second part of the stress σ_2 is that
due to nonequilibrium changes in the ensemble of tunnelling systems.
For this we use the Maxwell relation

$$-\frac{d(\delta\ell/\ell)}{dt} = \frac{1}{Y_2}\frac{d\sigma_2}{dt} \quad \frac{\sigma^2}{\eta_2} \tag{18}$$

where Y_2 is sometimes called "deviation" modulus and η_2 is the bulk
viscosity. With this notation the relaxation time τ is defined[11]
as $\tau = \eta_2/Y_2$.

In our case $d\sigma/dt = 0$, since the stress is kept constant after
application. Then the differential equation for $\delta\ell/\ell$ can easily
be solved and the heat generated by tensioning can be evaluated as

the difference between the total energy put into the system and the energy which is elastically stored. We obtain for the loss per volume Q_τ

$$Q = \frac{Y_2\sigma^2}{2T_1^2}(1 - e^{-\frac{2t_a}{\tau}}) \tag{19}$$

if $Y_2 \ll Y_1$, a condition which is always fulfilled. The index τ indicates that this expression is valid for a single value of τ. Exactly the same expression holds if we release the stress.

In order to calculate the heat observed in our experiment, we have to integrate over all tunnelling states with a relaxation time $t_s < \tau < t_a$. Using the distribution $P(\Delta,\lambda)$ introduced above we find

$$Q = \frac{\overline{PB_\Delta^2}\sigma^2}{2Y_1^2} \int_{\Delta=0}^{\infty} \int_{\lambda_{min}}^{\lambda_{max}} \frac{\Delta^2}{E^2} \frac{\partial f}{\partial E} (e^{-\frac{2t_s}{\tau}} - e^{-\frac{2t_a}{\tau}}) \, d\lambda d\Delta \tag{20}$$

Here we have replaced the "deviation" modulus Y_2 by $Y_2 =$

$\frac{\overline{PB_\Delta^2} \cdot \Delta^2}{E^2} \frac{\partial f}{\partial E}$, where $f = (1 + \exp(E/kT))^{-1}$. τ is given by equation 16.

The integral has to be evaluated numerically and is found to decrease only slowly with temperature in our temperature range. Of course the "deviation" modulus also appears in the expression for the ultrasonic absorption which is the dynamic analogue of our experiment.[9] From such experiments $\overline{PB_\Delta^2}/Y_1 = 2.5 \times 10^{-3}$ has been found for vitreous silica.[5] Putting in also the values for the other constants and carrying out the integration, we find $Q \simeq 10^{-3}$ erg/g for $\sigma = 1$ bar and $T = 2$ K. The agreement between this estimated value and the measured value of about 5×10^{-3} erg/g is surprisingly good, bearing in mind that the density of states which is used for the description of the dynamic behaviour of glasses in ultrasonic experiments at frequencies between 10 MHz and 10 GHz is extrapolated to our time scale which is of the order of seconds! In order to obtain this agreement, relatively large values for λ_{max} have to be put into equation 20. This shows that in the tunnelling model tunnel splittings as low as $\Delta_o/k = 10^{-5}$ K have to be considered, in contrast to previous work[12] discussing a lower limit of Δ_o^{min}/k of 16 mK. The longest relaxation times previously observed were[13,14] $\tau \leq 10^{-3}$ sec in measurements of the time dependence of specific heat at T < 1.5 K and in dielectric echo experiments[15] below 10 mK.

CONCLUSION

In conclusion we have established a new method of determining the Grüneisen parameter through the thermoelastic effect. This method promises to be very useful especially at low temperatures and in crystalline materials, where irreversible heating effects are not expected. Our experiments have confirmed that the Grüneisen parameter of vitreous silica decreases steadily at temperatures below 10 K in agreement with the tunnelling model. In addition we have found an unexpected irreversible heating effect due to tunnelling states with extremely long relaxation times which had not previously been observed.

ACKNOWLEDGEMENTS

We wish to thank K-H Heuschneider for participating in the early stage of this experiment and Prof R O Pohl and Prof J Jaeckle for stimulating discussions.

REFERENCES

1.   W A Phillips, J. Non-Cryst. Solids 31, 267 (1978).
2.   R C Zeller and R O Pohl, Phys. Rev. B4, 2029 (1971).
3.   S Hunklinger and M v Schickfus in: "The Physics of Amorphous
         Insulators", W A Phillips ed., Springer Verlag 1980 (in
         print).
4.   P W Anderson, B I Halperin and C Varma, Phil. Mag. 25, 1 (1972);
         W A Phillips, J. Low-Temp. Phys. 7, 351 (1972).
5.   S Hunklinger and W Arnold in: "Physical Acoustics", Vol. 12,
         eds. R N Thurston and W P Mason (1976) p. 155.
6.   K G Lyon, G L Salinger and C A Swenson, Phys. Rev. B19, 4231
         (1979).
7.   J L Black, Phys. Rev. B17, 2740 (1978).
8.   G K White, Phys. Rev. Lett. 34, 204 (1975).
9.   J Jäckle, Z. Physik 257, 212 (1972).
10.   M Reiner in: "Handbuch der Physik", Vol. VI, S. Flügge ed.,
         Springer Verlag, Berlin 1958, p. 434.
11.   R T Beyer and S V Letcher in: "Pure and Applied Physics"
         (H S W Massey and K A Brückner, eds.) Vol. 32, Physical
         Ultrasonics, Academic Press, New York, London (1969).
12.   J C Lasjaunias, R Maynard and M Vandorpe, J. Physique C6-973
         (1978).
13.   M T Loponen, R C Dynes, V Narayanamurti and J P Garno, Phys.
         Rev. Lett. 45, 457 (1980).
14.   M Meissner and K Spitzmann, to be published.
15.   L Bernard, L Piche, G Schumacher and J Joffrin, J. Low-Temp.
         Phys. 35, 411 (1979).

ULTRASONIC ABSORPTION IN POLYMETHYLMETHACRYLATE

AT LOW TEMPERATURES

G Federle and S Hunklinger

Max-Planck Institut für Festkörperforschung
Heisenbergstr. 1
D-7000 Stuttgart 80, Federal Republic of Germany

INTRODUCTION

In a series of experiments performed at low temperatures it has been demonstrated that amorphous solids exhibit thermal, acoustic and dielectric properties which are not observed in perfect crystals.[1] These anomalies are caused by a broad distribution of low-energy excitations which have their microscopic origin in structural re-arrangements of small groups of atoms via quantum mechanical tunnelling.[2] These tunnelling states interact strongly with phonons resulting in a strong ultrasonic absorption and a temperature dependent sound velocity even at temperatures of a few mK.[3]

In particular amorphous polymethylmethacrylate (PMMA) exhibits thermal properties at low temperatures (T < 4 K) which are characteristic for the glassy state.[4] Below 1 K the specific heat is considerably higher than that calculated from Debye's theory which accurately describes the behaviour of pure dielectric crystals. In contrast to expectation the heat capacity is not proportional to T^3. Measurements of the thermal conductivity indicate that even at the lowest temperatures (T < 0.1 K) the mean free path of thermal phonons is not limited by the size of the sample but by intrinsic scattering processes. Consequently the thermal conductivity of PMMA is not proportional to T^3 as in pure dielectric crystals but proportional to T^2 as in inorganic glasses.

Acoustic experiments have turned out to be valuable in the investigation of inorganic glasses. In polymers, only a few acoustic experiments at low temperatures have been reported. Apart from measurements in polystyrene[5] and two epoxide resins,[6] Brillouin scattering experiments down to 0.3 K have been carried out only in

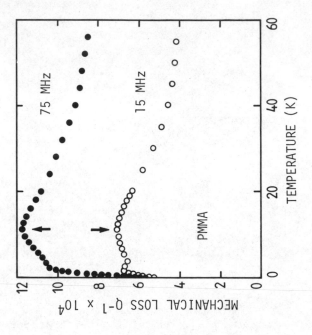

Fig. 2. Mechanical loss $Q^{-1} = \alpha v/\omega$ (where α is the absorption, v the sound velocity and ω the angular frequency) of longitudinal sound waves at 15 and 75 MHz. At 12 K an absorption maximum occurs as indicated by arrows.

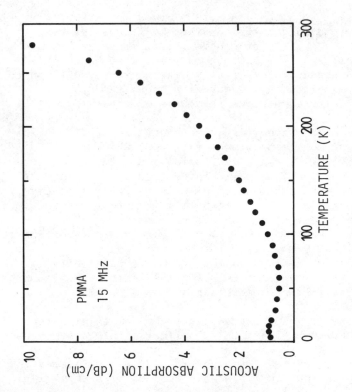

Fig. 1. Acoustic absorption of 15 MHz longitudinal sound waves in PMMA.

PMMA.[7] We have studied the absorption of longitudinal sound waves
at frequencies between 15 and 195 MHz and temperatures between
0.4 K and 300 K. Our sample was prepared from commercial PMMA and
exhibited not only similarities but also differences to the
absorption behaviour of inorganic and other organic glasses.

RESULTS

    In Figure 1 the attenuation at 15 MHz is plotted as a function
of temperature. On cooling the absorption steadily decreases until
60 K, where a shallow minimum is observed similar to that reported
for internal friction measurements.[8] Here we want to point out,
however, that at low frequencies this minimum was found in purified
samples only, whereas in our experiment we used commercial PMMA
without further treatment. At 12 K an absorption maximum is evident,
the position of which is frequency independent within the accuracy
of our measurement. (See arrows on Figure 2, where the loss factor
$Q^{-1} = \alpha v / \omega$ is plotted for 15 and 75 MHz.) At still lower temperatures, around 2 K a second peak occurs, which degenerates into a
shoulder at higher frequencies. The behaviour at the lowest
temperatures is shown in more detail in Figure 3. Here we have
plotted the reduced mechanical loss $Q^{-1}(T) - Q^{-1}(T = 0.4 \text{ K})$ for
15 and 195 MHz. The most interesting feature here is the linear
variation of the absorption with temperature below 1.5 K. A
similar result has been reported for polystyrene.[5] In the whole
temperature and frequency range the absorption $\alpha \propto \omega^n$, where
$1 < n < 2$.

THEORETICAL

    It is tempting to try to explain the low temperature properties
of glassy polymers in the same manner as in the case of inorganic
glasses. Therefore we presume that in PMMA tunnelling states also
exist which exhibit a broad distribution of their microscopic
parameters. In fact, such an assumption leads to a successful
description of specific heat,[4] thermal conductivity,[4] thermal
expansion[10] and previous acoustic measurements by Brillouin
scattering.[7] In this model it is assumed that small groups of
atoms are moving in double-well potentials, as shown diagrammatically in Figure 4. At low temperatures ($kT \ll V$, where V is the
barrier height) the "particle" is not able to surmount the barrier
in order to go from one site to the other, but quantum mechanical
tunnelling through the barrier can take place. Formally, such a
system is equivalent with a two-level system having an energy
splitting E which is determined by the asymmetry Δ in the depth of
the two wells and the overlap of the wave function of the tunnelling
"particle" at both sites:

$$E^2 = \Delta^2 + \Delta_o^2 \qquad\qquad (1)$$

To a good first approximation the tunnelling splitting is given by:

$$\Delta_o = n\Omega e^{-\lambda} \text{ with } \lambda^2 = 8mVd^2/n^2 \tag{2}$$

where Ω is of the order of the oscillator frequency of the "particle" in one well, and 2d the distance between the two minima.

Because of the random nature of the amorphous structure the parameters Δ and λ will not have well-defined values but rather a broad distribution. In the so-called "tunnelling model"[2] the distribution of the two parameters is assumed to be constant:

$$P(\Delta,\lambda)d\Delta d\lambda = \overline{P} d \Delta d \lambda \tag{3}$$

$\overline{P}$ is a constant and can be determined from specific heat measurements. An external strain (e) will vary the parameters of the tunnelling systems. In ultrasonic experiments this modulation is most likely due to a modulation of the asymmetry,[2,3] ie

$$\delta\Delta = B \cdot e \tag{4}$$

where B is the deformation potential and is of the order of 1 eV in inorganic glasses.[3] The effect of the modulation of Δ is twofold: firstly, it leads to a modulation of the energy splitting and secondly it induces a transition between the two energy states.

For a relatively low frequency the dominant interaction between an ultrasonic wave and the tunnelling states is expected to be a relaxation process.[10] Travelling through an ensemble of tunnelling stages, the sound wave disturbs the thermal equilibrium because of the modulation of the energy splittings by the external strain. The tunnelling systems try to re-establish thermal equilibrium via the emission and absorption of thermal phonons. The resulting relaxation absorption is given by:[11]

$$\alpha = \frac{4\overline{P}B^2}{\rho v^3 kT} \iint d\Delta d\lambda \frac{\Delta^2}{E^2} \frac{\partial f}{\partial E} \frac{\omega^2 \tau}{1 + \omega^2 \tau^2} \tag{5}$$

where ρ is the mass density, v the velocity of sound, and $f = (1 + \exp E/kT)^{-1}$. The relaxation time τ of the tunnelling states can easily be calculated if only one-phonon processes are taken into account:[11]

$$\tau^{-1} = \left(\frac{B^2}{v_\ell^5} + \frac{2B_t^2}{v_t^5} \right) \frac{\Delta_o^2 E}{2\pi\rho n^4} \cotn \frac{E}{2kT} \tag{6}$$

where B and B_t are the deformation potentials for longitudinal and transverse sound waves, respectively. The distribution of the parameters of the tunnelling systems also leads to a distribution of the relaxation times even for tunnelling systems of a well-defined energy splitting. The minimum value of the relaxation time τ_{min} is obtained for symmetric tunnelling states, since in this case Δ_o has its maximum value.

Although the integration of equation 5 can only be carried out numerically, in general two limiting cases can be calculated analytically:[10] at low temperatures the condition $\omega\tau_{min} \gg 1$ holds, leading to

$$\alpha = A \ \omega^o T^3 \tag{7}$$

where A is a constant. At higher temperature eventually $\omega\tau_{min} \ll 1$ will be fulfilled resulting in

$$\alpha = \frac{\pi\bar{P}B^2\omega}{2\rho v^3} \tag{8}$$

Although such a description holds perfectly well in fast ionic conductors[12] and is a good first approximation in inorganic glasses,[11] neither equation 7 nor equation 8 is adequate in the case of PMMA. In the following we try to modify the original tunnelling model in such a way that we obtain quantitative agreement.

DISCUSSION

Let us first discuss the absorption below 1 K where $\alpha \propto \omega^{1.25}$ T, according to our data in Figure 3. In the original tunnelling model a linear temperature dependence is only expected in the narrow temperature range where the transition from equation 7 to equation 8 occurs. Nevertheless the observed frequency dependence is close to that predicted by equation 8. Therefore we assume that the condition $\omega\tau_{min} \ll 1$ is really fulfilled above 0.4 K. In order to fit the temperature dependence, the constant distribution of equation 3 has to be replaced by a distribution depending linearly on the energy splitting, ie $P(E) = P_1E$, where P_1 is a constant. For this distribution and $\omega\tau_{min} \ll 1$ we find

$$\alpha = \frac{\pi P_1 B^2}{\rho v^3} \ kT \ ln2 \tag{9}$$

in qualitative agreement with the experiment. From equation 6 and the condition $\omega\tau_{min} \ll 1$ we can calculate a lower limit for the magnitude of the deformation potential B. Under the assumption[3] $B \simeq 2B_2$ we obtain $B > 0.6$ eV. This value is close to that known for inorganic glasses. We also can deduce an upper limit for P_1 by putting B into equation 9, ie we can roughly estimate the number of tunnelling states participating in this attenuation process.

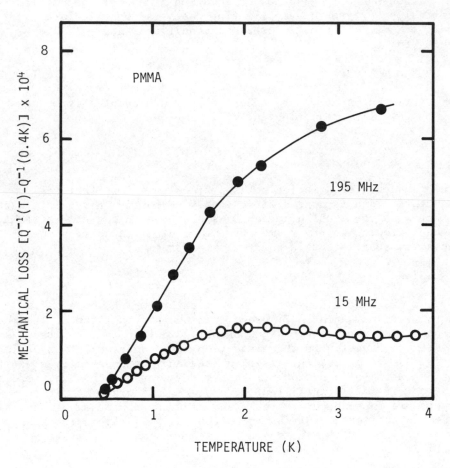

Fig 3. Reduced mechanical loss $Q^{-1}(T) - Q^{-1}(T = 0.4\ K)$ of longi-
        tudinal sound waves of 15 and 195 MHz. Note that the loss
        increases linearly with temperature below 1.5 K.

Using the data of Figure 3 we get $P_1 < 2 \cdot 10^{47}$ erg^{-2} cm^{-3}. This
is a surprisingly small fraction, approximately only 1% of the
density of states deduced from specific heat measurements.[4] Beyond
2 K the absorption mechanism seems to become abruptly inefficient.
This can only happen if the linearly increasing density of states
drops rapidly above a certain energy. Since in general tunnelling
states of E $\simeq$ 3 kT give the main contribution to the absorption
this discontinuity should occur around E/k = 6K.

Let us now discuss the absorption maximum observed at 12 K.
In general a maximum in the relaxation absorption is observed if
$\omega\tau$ = 1. Only in the case of an extreme wide distribution of relaxation times, as in the tunnelling model, this peak degenerates into
a constant absorption at higher temperatures. If, however, the peak
is determined by $\omega\tau$ = 1 and τ depends on temperature, then the
position of the peak shifts with frequency. For this reason we try
to explain the occurrence of the absorption in a different way.

From NMR measurements in polymers[13] it is known that in PMMA
the ester methylgroup is mobile down to temperatures of at least
4 K. It has been proposed that reorientation of this group occurs
via a hindered rotation. We believe that this structural unit is
also responsible for the acoustic attenuation. We assume that the
ester methyl group is moving in an asymmetric double well potential.
Such an asymmetry could be due to the presence of the carbonyl-
group in the side chain. Unfortunately detailed calculations are
not known to us. Since the shape of the potential is caused by a
well-defined structure we may assume that the parameters of this
potential are not distributed. In this case equation 5 can be
replaced by

$$\alpha = \frac{4NB^2}{\rho v^3} \frac{\partial f}{\partial E} \frac{\omega^2 \tau}{1 + \omega^2 \tau^2} \tag{10}$$

Here N is the number of relaxing centres per unit volume and therefore about 10^{22} cm^{-3} in our case. If τ is not temperature dependent
(as we will show below), then the factor $\partial f/\partial E$ determines the
temperature variation of the absorption. This will lead to a broad
maximum at T = 12 K for E/k = 18 K. From the magnitude of the
absorption measurement ($\omega\tau$ is of the order of unity in our frequency
range) we obtain B $\simeq$ 6 $\cdot$ 10^{-3} eV. Compared with the coupling
strength of the low-energy tunnelling states (where B > 0.6 eV was
found), this value is extremely small. This result is, however,
not surprising since the coupling between the rotational motion of
a side group and the main chain is expected to be weak. What
information can we obtain about the relaxation time τ? Firstly,
the temperature variation given by cotnE/2kT is in fact rather
small for kT < E and can be neglected in our crude considerations.
Secondly, the magnitude of τ can be estimated from the frequency
dependence of the absorption and consequently also the magnitude

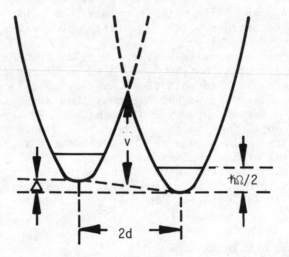

Fig 4. The double-well potential of the tunnelling model.

of Δ_0. We obtain 3 nsec for τ and a tunnel splitting corresponding to 2 K. Obviously the total energy splitting of 18 K is mainly due to the asymmetry of the potential.

SUMMARY

Our measurements show that the acoustic absorption of PMMA does not steadily decrease on cooling. At low temperatures two important features are found. Below 1 K the absorption varies linearly with temperature and can be explained after a small modification of the so-called "tunnelling model" which is well-known in the description of inorganic glasses. The microscopic nature of these states is, however, unknown. At 12 K an unexpected maximum is observed the position of which is frequency independent. We can explain this behaviour by assuming a high density of tunnelling states with well-defined asymmetries and tunnel splittings. These states couple weakly to the amorphous network and are most probably due to rotational tunnelling of the ester methyl group.

REFERENCES

1.   W A Phillips, J. Non-Cryst. Solids 31, 267 (1978).
2.   P W Anderson, B I Halperin and C Varma, Phil. Mag. 25, 1 (1972);
        W A Phillips, J. Low-Temp. Phys. 7, 351 (1972).
3.   S Hunklinger and W Arnold in: "Physical Acoustics", Vol. 12,
        eds. R N Thurston and W P Mason (1976), p. 155.
4.   R B Stephens, Phys. Rev. B8, 2896 (1973).
5.   J Y Duquesne and G Bellessa, J. Physique Lett. 40, L193 (1979).
6.   D S Matsumoto, C L Reynolds and A C Anderson, Phys. Rev. B
        19, 4277 (1979).
7.   R Vacher, J Pelous, H Sussner, M Schmidt and S Hunklinger,
        Solid State Communications, to be published.
8.   A E Woodward, J. Polymer Science C, 14, 89 (1966).
9.   K G Lyon, G L Salinger and C A Swenson, Phys. Rev. B19, 4231
        (1979).
10.   J Jäckle, Z. Physik 257, 212 (1972).
11.   J Jäckle, L Piche, W Arnold and S Hunklinger, J. Non-Cryst.
        Solids 20, 365 (1976).
12.   P Doussineau, R G Leisure, A Levelut and J Y Prieur, J. Physique
        Lett. 41, L65 (1980).
13.   K M Sinnott, J. Polymer Science 42, 3 (1960);
        G Filipovich, C D Knutson and D M Spitzer, Jr., Polymer Lett.
        3, 1065 (1965).

RADIATION DAMAGE IN THIN SHEET FIBREGLASS INSULATORS

Elena A Erez and Herbert Becker

MIT Plasma Fusion Center
Cambridge
Massachusetts

INTRODUCTION

The term "radiation damage", as applied to structural behaviour, can be defined as the reduction in load-carrying ability resulting from exposure to radiation. It has been observed that the radiation induced loss in strength of a material could depend upon the type of load to be resisted.[1-4] The data in this paper indicate that the structural configuration could be of major importance.

The research was part of an investigation into the properties of materials for an ignition test reactor (ITR) which is being designed to study the physics of fusion ignition. The reactor magnet consists of large flat plates of copper/steel composite separated by thin insulator sheets. The insulator must survive 10,000 cycles of 20 ksi (140 MPa) pulsed pressure, 1.2 ksi (8.4 MPa) pulsed interlaminar shear stress and a lifetime radiation fluence of 10^{20} n/cm^2. The pulses could occur at 30 minute intervals. Furthermore, each cycle would start at 77 K and end near 150 to 200 K. In addition, the insulator must have a coefficient of friction of at least 0.30.

Most of the existing test data on insulator radiation survivability have been obtained from static flexural and compression tests of rods.[2-8] It is likely that these results do not apply to thin sheets under cyclic compressive load. Preliminary studies on unirradiated insulators indicated that a 1/2-millimetre-thick fibreglass composite with an organic matrix might withstand the ITR environment. Consequently, a programme of irradiation and test was carried out to explore that possibility.

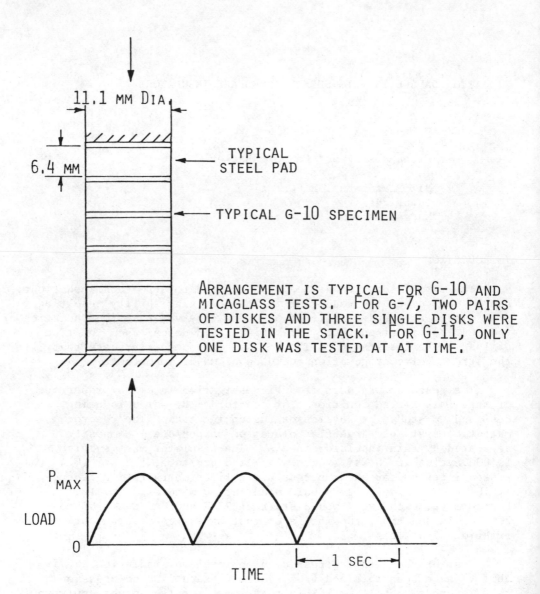

Figure 1. Test fixture schematic and loading cycle

RATIONALE

The failure mode in a compressed thin sheet of brittle material is different from that of a rod. The stress distribution in a rod is uniaxial and failure usually occurs on the familiar diagonal shear plane, more or less at 45 degrees to the rod axis. The thin sheet also would be under uniaxial compression if a pure pressure were to be applied. However, the insulators on the ITR are compressed between large flat plates. As a result, there is friction-induced restraint in the plane of the sheet similar to the behaviour studied by Bridgman.[9] The diagonal shear failure planes cannot form easily. Failure can occur only by crushing. Observations reveal that the insulator specimens are reduced to powder by extensive compressive cycling, in support of that hypothesis.

Failure of G-10 and similar grp materials may begin by crushing at the intersections of the cloth warp and fill fibres. Any tendency for the cloth to spread would be resisted by friction from the metal plates retarding breakage until the fibres begin to crush between intersections. The matrix material would help to support the fibres during that process. The onset of failure has been observed to be accompanied by rapid degradation of stiffness.

Development of a quantitative theoretical explanation would require a more extensive study. Until that time, the above rationale has been adopted as part of the basis for believing that materials like grp can withstand the ITR fluence at the design compression stress for the required number of cycles.

FAILURE CRITERION

It is a simple matter to observe failure in compressed brittle rods. A break occurs and the testing machine load drops suddenly toward zero. In thin sheets, however, the failure process is not so obvious. This is particularly true of fatigue loading.

It was noticed, during exploratory tests on unirradiated specimens, that the stiffness appeared to increase by a few percent up to approximately 5000 to 10,000 cycles after which the stiffness reduced relatively rapidly up to 100,000 cycles. The same phenomenology was observed during the INEL tests on irradiated specimens except that the degradation in stiffness occurred in a few hundred cycles. Subsequent examination showed that at least one disk in a stack of five had been reduced to powder.

It was decided to define failure in thin sheets as the rapid reduction of stiffness. The relevant data were chosen as the stress level and the number of cycles at which that rapid reduction occurred.

INITIAL TESTS

Experiments were carried out with sheets of fibre reinforced plastics and one common inorganic electrical insulator. Unirradiated specimens of G-7, G-10 and micaglass were subjected to compression fatigue at RT. Both G-7 and G-10 are commercial E-glass reinforced plastics. The matrix system of G-7 is a silicone resin while that of G-10 is an epoxide resin. The test fixture and loading scheme are shown in Fig 1.

The initial test results appear in Table 1. The grp survived pressures twice as high as in ITR for the required 10,000 cycles. The micaglass, however, did not survive under pressures 50 percent greater than in ITR. The 1 Hz frequency was chosen as a practical compromise between the low ITR cycle and the need for shorter test times to collect data from several samples.

Additional tests (Table 1) were performed on unirradiated specimens selected from the composite formulations described below. These results also indicated high survivability potential.

TESTS ON IRRADIATED SPECIMENS

INEL Tests

Disks were cut from thin sheets of G-7, G-10 and G-11 CR*. They were irradiated in the Advanced Test Reactor at Idaho National Engineering Laboratory. The radiant flux was calculated from a standard code used at INEL and is stated to be within 20 percent of actual values. The total fluence was 1.6×10^{19} n/cm^2 for neutron energies greater than 0.1 MeV, 10^{20} n/cm^2 for the total neutron spectrum and 3.8×10^{11} rads of gamma radiation. That dose is somewhat higher than the fluence expected in ITR.

The specimen temperature was reported to be 120°F. All specimens were found to be highly radioactive after months of cooldown. Consequently, testing was conducted in a hot cell.

The compression fatigue tests were conducted in the same manner as for the unirradiated samples (Fig 1). The results appear in Table 2. In addition the G-10 data are plotted on the graph of Fig 2. All tests were stopped arbitrarily at 200,000 cycles if no failure had been observed.

It is clear that the observed strengths are much greater than reported previously for rods irradiated at 4.9 K^6 for which G-10 CR static compression values of about 69 MPa were obtained. The

*Diglycidyl ether of bisphenol A reinforced by E-glass.

Table 1. Results of Compression Fatigue Tests of
Unirradiated Samples at R.T.
(5 specimens of each type tested in stack
shown in Figure 1).

INITIAL TESTS

Material	Thickness (mm)	Max. Applied Stress (MPa)	Number of Cycles
G-7	0.3	207	10.000 S
		276	10,000 S
		207	100,000 F
G-10	0.50	310	60,000 S
Mica-Glass	0.50	207	10,000 F

S = Survived, F = Failed

ADDITIONAL TESTS

Material		Thick (mm)	Max.Applied Stress (MPa)	Number of Cycles
Matrix System	Reinf.			
Kerimid 601	S	0.50		
TGPAP & DCA	S2	0.50		
DGEBA & DDS	S2	0.50	All 310	All 60,000
TGPAP & DDM	E	0.46		
TGPAP & DDM	S2	0.48		
TGPAP & DDS	S2	0.50		

Tests were halted arbitrarily at indicated number of cycles.
All specimens survived.

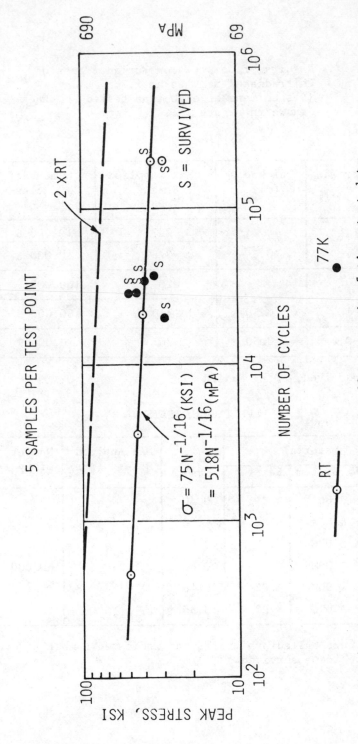

Figure 2. Irradiated G-10 compression fatigue test data

INEL results also exceed the ITR requirements. The stress level of
345 MPa is more than twice the ITR requirement. Furthermore,
200,000 cycles corresponds to 20 times the required life.

If it is assumed that the low temperature fatigue strength is
twice the RT value, which matches the ratio for static ultimate
compression of G-10 rods, then the 77K fatigue curve would be as
shown on Fig 2. The observed survivability of the 77K specimens
is consistent with that curve.

MIT Tests

The INEL tests were considered to support the rationale that
G-10 might survive the ITR environment. It was important to obtain
independent data as a further check. It also was decided to broaden
the scope of the programme by including other candidate insulators.

A searth of the literature showed that epoxide and polyimide
resins with fibreglass reinforcement could be considered as candidate
insulation for ITR. Of the epoxide resins, glycidyl amines were
concluded to be more radiation resistant.[2,3] The aromatic amine
hardeners lead to resin systems which appear to be more stable under
radiation than do anhydride hardeners.[3] According to reference 2,
glycidyl amine and glycidyl ether resins are best when combined with
an anhydride hardener whereas novolac is best when combined with an
aromatic amine.

Most radiation tests have been carried out with commercial
laminates with E-glass reinforcement.[5,6] Because of greater purity,
S-glass may provide a more useful reinforcement than E-glass for
radiation resistant insulators. In reference 1 it is shown that
boron-free glass begins to show damage under 10^{16} n/cm^2 while
quartz shows no damage up to 10^{21} n/cm^2 (E > 0.1 MeV).

Sheets of composite were prepared from two epoxide resins
(glycidyl amine and glycidyl ether) mixed with aromatic amine and
anhydride hardeners in combination with E-glass, S-glass and quartz
fibre reinforcement. Polyimide resins were also employed with these
three types of reinforcements. The components are shown in Table 3.

Twenty-eight types of specimens (Table 4) of various thicknesses were evaluated for residual radioactivity. They were
irradiated in the MIT Reactor for 96 hours. The total fluence was
1.4×10^{18} n/cm^2 and 5×10^9 rads of gamma radiation.

The activation of each specimen was measured 258, 330 and 450
hours after irradiation. The S-glass composites were much less
active than E-glass composites. This agrees with the INEL observations regarding the high residual activity of E-glass composites.

Table 2. Results of INEL Compression
Fatigue Tests of Irradiated Insulators

For all Specimens D = 11.1 mm
(See Figure 1 for test arrangement)

Material	Thickness (mm)	Temperature	Max.Applied Stress (MPa)	Number of Cycles
G-7	0.30	RT	207	10,000 F*
G-11	4.00	RT	207	10,000 F
G-10	0.50	RT	207	200,000 S
			241	200,000 S
			276	21,900 F
			310	3,570 F
			345	460 F
		77K	207	20,000 S
			241	40,000 S
			276	36,000 S
			310	30,000 S
			345	30,000 S

*Paired disks broke, singles survived

Table 3. Common Resins, Hardeners and Reinforcements
Used in Insulators

Resins

Designation	Classification	Chemical Name	Trade Name
TGPAP	Epoxide	Triglycidyl p-amino phenol	Ciba 0500
DGEBA	Epoxide	Diglycidyl ether of Bisphenol A	Ciba 6010
KERIMID 601	Polyimide	Bis-maleimide amine	Rhodia Merimid 601

Hardeners

DDM	Aromatic	Diaminodiphenyl methane	Ciba 972
DDS	Aromatic	Diaminodiphenyl sulfone	Ciba Eporal
OCA*	Anhydride	Proprietary, Owens-Corning	---

*MIT Designation

Woven Fabric Reinforcement
(All specimens contained approximately 70 percent by volume)

Material Designation	Weave Style	Finish	Manufacturer
E-glass	181	A-1100	Clark-Schwebel
E1-glass	181	P-283B	Owens-Corning
S-glass	181	901	Owens-Corning
S2-glass	6581	GB-770B	Burlington
Quartz	527	A-1100	J P Stevens

Table 4.   Specimens Irradiated in M.I.T. Reactor

| Specimen Type | Material | | Thickness mm |
	Matrix System	Reinforcement	
1	DGEBA + OCA	E	0.50
2		E1	0.50
3		S	0.56
4		S	2.79
5		S2	0.50
6		None	3.05
7	TGPAP + OCA	E	0.50
8		E1	0.50
9		S	0.56
10		S2	0.50
11		None	3.18
12	TGPAP + DDM	E	0.46
13		E1	0.50
14		S2	0.48
15		Quartz	0.43
16	TGPAP + DDS	E	0.46
17		S2	0.50
18	DGEBA + DDS	E	0.46
19		S2	0.50
20	Kerimid 601[1]	E	0.46
21		S	0.50
22		Quartz	0.50
23	Polyimide NR150B2	Quartz	0.50
24	PNE +APF[2]	E	0.50
25	G-7		0.30
26	G-11 CR		0.50
27			4.00
28	G-10		0.50

The composites in Table 4 were irradiated in the MIT reactor to 2.3 x 10^{10} rads of gamma radiation, 1.06 x 10^{19} n/cm^2 at E > 1 MeV and 2.16 x 10^{19} total n/cm^2. This is roughly equivalent to 1/5 of the ITR fluence. Compression testing is in progress at RT. Some of the early results appear in Table 5. As can be seen, the strengths exceed the ITR requirements. Furthermore, they are higher than for the INEL fluence.

Friction Tests

One of the first insulation candidates was mica paper which was considered to be free of damage at the ITR fluence level. The potential use was tentatively ruled out partially as a result of tests that revealed a coefficient of friction of 0.033 to 0.049 at RT and 0.079 to 0.092 at 77 K. The relatively poor showing in the preliminary compression tests (Table 1) was another reason.

G-10, on the other hand, exhibited a minimum coefficient of 0.33 between RT and 4.2 K.[10]

The above results were obtained on unirradiated materials. If irradiation reduces structural strength then it might also reduce frictional resistance. Test data are required in this area.

CONCLUSIONS

Evidence has been obtained at RT to support the rationale that thin sheet grp can withstand the ITR radiation and compression loading environment. It remains to conduct combined interlaminar shear and compression tests during irradiation at 77 K before the survivability of grp can be established reliably for use in ITR.

FUTURE TESTING

Compression testing will continue at RT and at 77 K on the remainder of the large variety of specimens irradiated at MIT.

A fixture has been designed for exploratory tests under combined normal compression in conjunction with interlaminar frictional shear (C-S tests). Static load and fatigue experiments will be conducted in that fixture at INEL on irradiated grp specimens at RT and 77 K. MIT also plans C-S tests on specimens already irradiated in the MIT reactor.

Inpile compression fatigue testing at 77 K is now being planned. A friction test programme also is being designed to obtain data at RT and 77 K on irradiated specimens.

Table 5.   Results of M.I.T. Compression
Tests of Irradiated Insulators
(5 specimens of each type tested in
stack shown in Figure 1).

Material		Thickness (mm)	Max.Applied Stress (MPa)	Number of Cycles
Matrix System	Reinf.			
Kerimid 601	S	0.50		10,000
DGEBA + OCA	S	0.56		30,000
DGEBA + OCA	S2	0.50	All 310	30,000
TGPAP + OCA	S	0.56		30,000
TGPAP + OCA	S2	0.50		30,000

Tests arbitrarily halted at indicated number of cycles.   All
specimens survived.

ACKNOWLEDGEMENTS

    The materials for this programme were donated by the Spaulding
Fibre Company and by the Owens-Corning Fibreglass Corporation. The
authors appreciate the help of J Benzinger of Spaulding and
J Olinger of Owens-Corning for their assistance and for numerous
enlightening discussions.

    Thanks are due to R E Schmunk (of EG&G) for directing the
irradiation and testing at INEL and to G Imel (also of EG&G) for
his assistance in developing the test programme. They both offered
numerous helpful suggestions.

    W Fecych is to be commended for his organisation and direction
of the MIT irradiation programme. C-H Tong is to be congratulated
for performing the MIT compression testing programme so well at
short notice.

    Appreciation is expressed to Y Iwasa and R Kensley for the
friction testing of micaglass.

REFERENCES

1.   Anon., "Radiation Effect Design Handbook" Section 3.
          Electrical Insulation Materials and Capacitors, NASA CR
          1787, July 1971.
2.   E Laurant, "Radiation Damage Test on Epoxies for Coil
          Insulation", NAL, EN-110, July 1969.
3.   D Evans, J T Morgan, R Sheldon, G B Stapleton, "Post-
          Irradiation Mechanical Properties of Epoxy Resin/Glass
          Composites", RHEL/R200, Chilton, Berkshire, England, 1970.
4.   M H Van de Voorde, "Selection Guide to Organic Materials for
          Nuclear Engineering", CERN 72-7, May 1972.
5.   G R Imel, P V Kelsey and E H Ottewitte, "The Effects of
          Radiation on TFTR Coil Materials", 1st Conference on
          Fusion Reactor Materials, January 1979.
6.   R R Coltman, Jr., C A Klabunde, R M Kernohan and C J Long,
          "Radiation Effects on Organic Insulators for Superconducting
          Magnets", ORNL/TM-7077, November 1979.
7.   H Brechna, "Effect of Nuclear Radiation on Organic Materials;
          Specifically Magnet Insulations in High-Energy Accelera-
          tors", SLAC Report No 40, 1965.
8.   K Shiraishi, Ed. "Report of Group 6 Materials", IAEA Workshop
          on INTOR, June 1979.
9.   P W Bridgman, "The Physics of High Pressure", G Bell, London,
          1931.
10.   R S Kensley, "An Investigation of Frictional Properties of
          Metal-Insulated Surfaces at Cryogenic Temperatures", MIT
          MS Thesis, June 1979.

EPOXIDE RESINS FOR USE AT LOW TEMPERATURES

D Evans and J T Morgan

Rutherford and Appleton Laboratories
Chilton
Didcot, Oxon OX11 0QX

INTRODUCTION

Materials based on epoxide resins are frequently used in the manufacture of nuclear physics apparatus, such as bubble chambers, superconducting magnets and ancillary equipment. In such applications, these materials are required to operate under high stress at temperatures below 20 Kelvin.

Epoxide resins, in common with other polymers, are much less resilient at low temperatures and additionally, compared with metals, have large thermal expansion coefficients and low thermal conductivities.

In formulating epoxide resins for low temperature applications, two distinct approaches are possible. Firstly the incorporation of fillers in amounts which will reduce the thermal expansivity to comparatively low levels. Secondly, and the one with which this report is primarily concerned, is the incorporation into the polymerised structure of certain types of molecular species which confer a degree of ductility to the resin. This approach limits the stresses induced by differential thermal contraction because the material does not become brittle until some point below room temperature.

The first approach, ie the use of non reinforcing fillers, is satisfactory only for low stress applications because of the deleterious effect on mechanical properties caused by the incorporation of inert fillers.

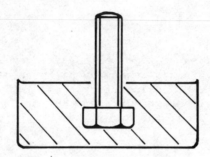

Fig. 1.   Thermal shock specimen.

The second approach is more appropriate to applications where the high viscosity of filled systems precludes their use (eg for the impregnation of closely wound coils) or where laminating techniques using glass or other fibres are employed.

For high stress applications, the incorporation of reinforcing fibres in the form of fabrics or rovings is essential but due regard must be given to problems which might arise from the anisotropicity of such composites.[1]

PHYSICAL PROPERTIES OF SELECTED RESIN SYSTEMS

The number of epoxide resin systems that could be formulated is almost infinite but so far only a few unfilled systems have been identified as suitable for use at very low temperature. For this work we used a simple screening test to permit the rapid evaluation of many epoxide resin systems, with a subsequent more detailed evaluation of those which performed well in this test. In this manner, several hundred resin systems were investigated including such variables as hardener ratios and blends, diluents, flexibilisers and modifiers, but considering only one epoxide resin, namely diglycidyl ether of Bisphenol A. Future work will include a programme of resin evaluation using only a limited range of hardeners.

Initial Screening Test

A number of thermal shock and resin shrinkage tests have been reported but in the main these tests have proved to be of little quantitative value. It was therefore decided to adopt a simple qualitative test specimen comprising a resin block containing a metallic insert, shown in figure 1.

It consists of a resin block 73 mm diameter by 25 mm thick containing a 10 mm by 50 mm hexagonal headed brass bolt, located axially with the threaded end protruding and having approximately 3 mm between the encapsulated head and surface of the specimen.

These specimens were plunged into liquid nitrogen, removed when completely cold, allowed to warm to room temperature and examined for signs of cracking. Uncracked specimens were re-cycled, making the procedure into a low frequency fatigue test.

Resin blocks without the brass bolt insert were similarly tested.

Viscosity and Gel Times

These properties of the mixed resins were determined using a Brookfield viscometer and a commercial gel point timer respectively, the gel time being recorded for batches of approximately 200 grams.

Table 1

FORMULATION NUMBER	COMPOSITION		PARTS BY WEIGHT
1	Diglycidyl Ether of Bisphenol A (DGEBA)	(a)	100
	Methyl 'Nadic' Anhydride		80
	Accelerator	(b)	1
53	DGEBA		100
	Polyoxypropylene diamine (Mwt 230)	(c)	40
71	DGEBA		100
	Polyoxypropylene diamine (Mwt 400)		57
	Cycloaliphatic amine	(d)	10
79	DGEBA		60
	Epoxide resin based on polypropylene glycol	(e)	40
	Polyoxypropylene diamine (Mwt 230)		35
93	DGEBA		100
	Diamino diphenyl methane solution	(f)	50
	Accelerator		1
222	DGEBA		100
	Polyoxypropylene diamine (Mwt 2000)		37
	Diamino diphenyl methane		20
	N-methyl-2-pyrrolidone		7.5
226	DGEBA		100
	Polyoxypropylene diamine (Mwt 230)		35
	Polyoxyethylene diamine (Mwt 600)	(g)	20
227	DGEBA		100
	Polyoxypropylene diamine (Mwt 2000)		37
	Polyoxypropylene diamine (Mwt 400)		15
	Polyoxypropylene diamine (Mwt 230)		20
228	DGEBA		100
	Polyoxypropylene diamine (Mwt 400)		55
	Diamino diphenyl methane solution	(f)	15
	Accelerator		1
229	DGEBA		100
	Flexibilising amine	(h)	27
	Polyoxypropylene diamine (Mwt 2000)		80

NOTES ON TABLE 1

Typical Manufactured Products

    (a) Ciba Geigy MY 740.
    (b) Ciba Geigy HY 906.
    (c) Texaco D230.
    (d) Schering Chemicals-Euredur 40.
    (e) Ciba Geigy CY 208.
    (f) Ciba Geigy HY 219.
    (g) Texaco ED 600.
    (h) Ciba Geigy XD 716.

The resins were cured for at least 16 hours at room temperature followed by 24 hours at 60°C, with the exception of formulation number 1 which was cured for 12 hours at 120°C.

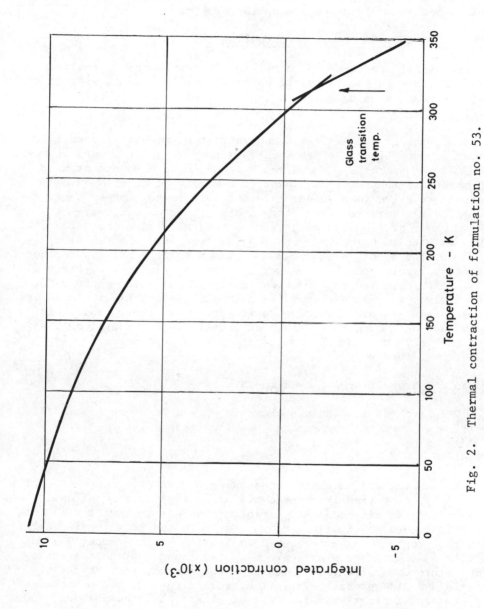

Fig. 2. Thermal contraction of formulation no. 53.

Hardness

This was determined using a Barcol hand held hardness tester, the reading being taken 10 seconds after application of the instrument to a flat moulded surface on the specimen.

Determination of Contraction on Cooling

For small specimens, a commercial Thermomechanical Analyser (TMA) was used to determine thermal contraction down to approximately 80K, the results being extrapolated to liquid nitrogen temperature (77K).

For measuring the thermal contraction of materials down to the temperature of liquid helium (4.2K), a purpose built apparatus was used. This apparatus was designed to accommodate specimens up to 100 mm in length and 25 mm diameter, dimensional changes being recorded relative to a copper reference specimen. Larger specimens and mouldings were measured using a dial gauge as follows:- the specimen was set in an apparatus with the ball ended foot of a dial gauge located in a small depression machined into the end of the specimen, the opposite end of the specimen being similarly located against a reference point. After zeroing the dial gauge, the specimen was removed and immersed in liquid nitrogen. When cold, the specimen was removed from the nitrogen and quickly replaced in the apparatus in order to measure its change in dimension. This technique does not give readings at intermediate temperatures, as does the TMA, but measures the total integrated contraction from room temperature to 77K.

Mechanical Properties at Low Temperatures

Strength and modulus measurements were made using a specially modified Instron testing machine previously described.[2]

MATERIALS AND RESULTS

Because of its excellent processing characteristics, especially for impregnating tightly wound coils or fabrics, a conventional anhydride cured Bisphenol A resin has found use in a wide range of low temperature applications.[3] This is referred to as resin number 1 in this report. An amine cured laminating resin (number 93) capable of room temperature cure has also found application in low temperature environments.[4] Both of these resins are inherently brittle and did not pass our initial screening test but nonetheless perform satisfactorily in the form of glass fabric composites, and are included in this report.

In the search for materials having improved performance at low temperatures a large number of epoxide resin systems was

TABLE 2

RUTHERFORD FORMULATION NO	1	53	71	79	222
Mixed Viscosity at R.T. PaS	1.3	0.8	0.63	1.4	3.4
Gel Time Minutes R.T.	7400	610	990	650	6900
Gel Time Minutes 40°C	3100	130	220	200	2080
Gel Time Minutes 60°C	750	40	50	45	750
Barcol Hardness	80.8	74.0	63.5	65.5	66.0
Thermal Shock: Cycles prior to Failure	0	2	>25	22	>25
298K $\int$ αdt 77K	0.011	0.009	0.010	0.011	0.010
298K $\int$ αdt 4.2K	0.012	0.011	0.012		
Impact Strength[1] R.T. Nm/mm of notch	0.014	0.026	0.02	0.017	0.016
Impact Strength[1] 80K Nm/mm of notch	0.010	0.010	0.010	0.008	0.010
Flexural Modulus R.T. GNm^{-2}	3.8	3.0	2.7	2.6	1.5
Flexural Modulus 77K GNm^{-2}	6.9	7.5	7.2	7.6	7.4
Flexural Modulus 4.2K GNm^{-2}	7.5	7.6	7.6	7.8	7.7
Flexural Strength R.T. MNm^{-2}	140	95*	85*	89*	56*
Flexural Strength 77K MNm^{-2}	176	225	279	205	220
Flexural Strength 4.2K MNm^{-2}	151	216	203	226	188

TABLE 2 CONTD.

RUTHERFORD FORMULATION NO		93	226	227	228	229
Mixed Viscosity at R.T. PaS		1.5	0.60	1.0	0.54	1.4
Gel Time Minutes	R.T.	880	770	1200	990	2200
	40°C	200	180	240	150	360
	60°C	80	40		40	
Barcol Hardness		72.5	63.5	45.0	57.0	0
Thermal Shock: Cycles prior to Failure		0	7	>25	3	>15
$298K \int \propto dt$ 77K		0.010	0.010		0.009	0.013
$298K \int \propto dt$ 77K						
Impact Strength[1] Nm/mm of notch	R.T.	0.012	0.015	0.041	0.014	0.045
	80K	0.007		0.015	0.011	0.012
Flexural Modulus GNm^{-2}	R.T.	2.7	2.1	0.7	1.8	0.05
	77K	7.9	7.8	7.3	7.1	6.8
	4.2K	8.0	7.8		7.6	
Flexural Strength NMm^{-2}	R.T.	98	74*	32*	50*	2.8*
	77K	200	247	160	240	291
	4.2K	173	226		209	

1 Hounsfield   * Flexural Yield Strength (ASTM D790-69)

examined but the selection was mainly limited to those having a low
mixed viscosity and therefore suitable for impregnation purposes
or for the incorporation of high loadings of inert particulate
fillers.

Whilst it is appreciated that the curing cycle may influence
the properties of resins, this variable was largely ignored, various
standardised cure times and temperatures being adopted to limit the
size of the test programme. However, in our experience, undercure
is to be avoided because resins which are not fully cured have been
found to exhibit poor thermal shock resistance.

The physical properties for selected resin systems are given
in tables 1 and 2. A wide range of hardeners, flexibilisers,
plasticisers and other additives were included in the thermal shock
tests, and in summary it may be stated that:-

a) Non reactive flexibilisers, most of which increase the
   toughness of resins at normal temperatures offer no such
   improvement at very low temperatures.

b) Reactive diluents, such as mono glycidyl compounds,
   increase the tendency of resins to crack on cooling.

c) The incorporation of elastomeric modifiers* in anhydride
   cured resins[5] offers only a slight improvement to the
   cracking behaviour at low temperatures, with an associated
   deterioration in processing characteristics.

d) One group of hardeners/modifiers, namely diamines based on
   linear polyethers having molecular weights in range of
   200 - 2000, were by far the most effective in reducing the
   tendency of resins to crack in the thermal shock test.

A series of tests was carried out in which the amine hardener
of a laminating resin system was gradually replaced by a diamine
based on polypropylene oxide (Mwt approximately 2000). The effect
on properties is clearly shown in table 3. The material becomes
softer and more flexible at room temperature as the proportion of
modifier is increased but its resistance to thermal shock is a
maximum at about 40 parts of modifier (formulation number 105).

Apart from the laminating resins (numbers 1 and 93) none of
the materials broke on flexural testing at room temperature, the
quoted strength values being derived from the maximum loads.
Although the moduli at room temperature of the materials show
considerable differences, there is much less variation between the
values at low temperature.

_____
*Such as carboxyl terminated butadiene/acrylonitrile copolymers.

TABLE 3

RUTHERFORD FORMULATION NO		93	102	103	104	105	106	107
COMPOSITION	DGEBA	100	100	100	100	100	100	100
	HY219[1]	50	47	45	42.5	40	37.5	35
	POPDA 2000[2]	0	10	20	30	40	50	60
	Accelerator	1	1	1	1	1	1	1
Mixed Viscosity at R.T. PaS		1.5	1.8	2.0	2.2	2.3	2.3	2.4
Gel Time Minutes at R.T.		880	2600	3200	3400	3500	5000	5000
Barcol Hardness		72.5	70.5	66.3	57	45.7	33.2	13.4
Thermal Shock cycles prior to failure		0	0	0	1	5	3	2
$298K \int \propto dt$ 77K		0.010					0.012	0.013
Impact Strength Nm/mm of notch	R.T.	0.013	0.019	0.021	0.026	0.033	0.035	0.037
	80K	0.007	0.008	0.009	0.011	0.012	0.012	0.014
Flexural Modulus GNm^{-2}	R.T.	2.7	2.3	1.5	1.2	0.7	0.09	0.04
	77K					7.2		
	4.2K							
Flexural Strength MNm^{-2}	R.T.	98	89	55*		26*		2.3*
	77K					237		
	4.2K							

1 Ciba Geigy HY219　　2 Texaco D2000　　*Flexural Yield Strength

TABLE 4. IMPREGNATING CHARACTERISTICS OF SELECTED EPOXIDE RESIN SYSTEM.PENETRATION (in mm) INTO STANDARD COLUMN OF 'BALLOTINI'.

FORMULATION NUMBER / TEMP. ^{o}C	53	71	222	1	1*
20	112	112	80	–	–
30	–	158	108	–	–
40	126	146	131	170	255
50	123	135	143	172	260
60	120	132	150	155	280
70	–	–	145	134	244
80	–	–	–	127	178

* 0.5 phr accelerator content

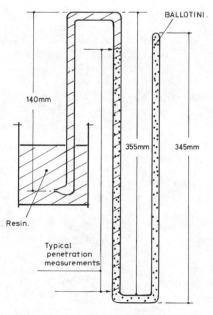

BALLOTINI.

140mm

355mm   345mm

Resin.

Typical
penetration
measurements

Fig. 3. Pyrex "U" tubes used in resin penetration tests

Figure 2 shows the dimensional changes versus temperature for one of the resin systems from room temperature to 4.2K and from room temperature to 350K.

The impregnating properties of some of these formulations have been characterised using a test developed at the Rutherford Laboratory. This test measures the maximum penetration of the resin into a standard column of tiny glass spheres (Ballotini) under vacuum conditions at various temperatures. Typical results are shown in table 4.

CONCLUSION

A number of unfilled epoxide resin systems offering improved resistance to thermal shock have been developed and characterised. However, the low temperature physical properties of cured resin systems are little changed by formulation variables. It is during cool down from room temperature that these newly developed resin systems demonstrate their ability to absorb the strains induced by differential thermal contraction.

The new systems such as formulations numbered 71 or 222 demonstrate excellent resistance to thermal shock and have impregnation characteristics suitable for the preparation of fibre reinforced composites by vacuum impregnation or by wet lay up procedures. Their improved resistance to thermal shock allows them to be used in situations where some inhomogeneity of reinforcement is unavoidable or where large temperature gradients may be present.

Further work is planned to assess the performance of these materials when used with fillers and in laminates in fatigue situations.

It should be re-stated that polymers used in low temperature applications should contain, wherever possible, reinforcement in the form of glass fabrics, rovings or chopped strands.

ACKNOWLEDGEMENTS

The authors are indebted to colleagues in the Chemical Technology Group, especially to R M Luckock and G W Hall, for the preparation and testing of specimens.

REFERENCES

1. D Evans and J T Morgan - 14th International Conference on
        Reinforced Plastics, Paris, 28-29 March 1979.
2. D Evans, C E Micklewright, R Sheldon and G B Stapleton -
        Mechanical Testing at 4.2K, Rutherford Laboratory Report
        RPP/E10.

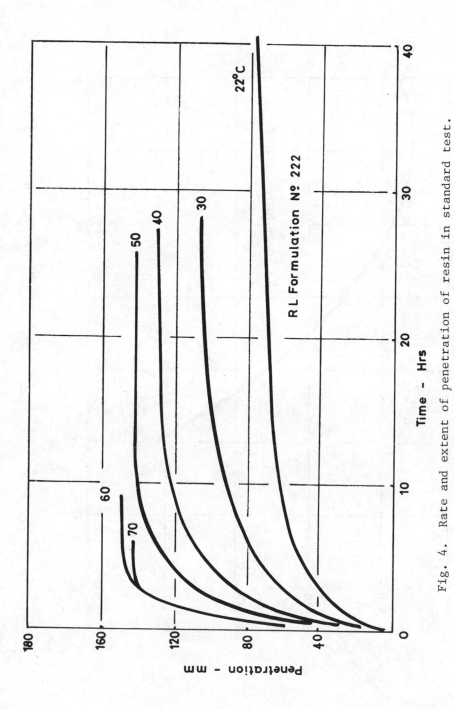

Fig. 4. Rate and extent of penetration of resin in standard test.

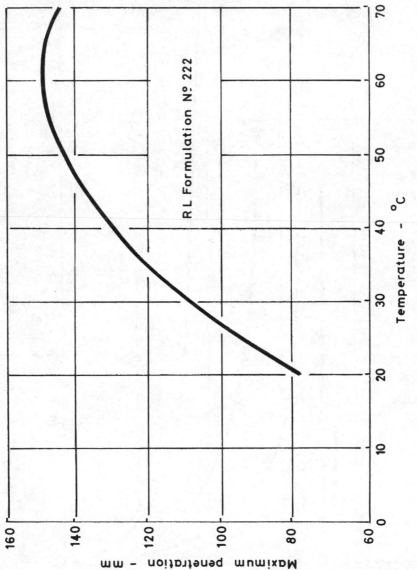

Fig. 5.   Extent of penetration of resin as a function of temperature.

3.   D Evans, J U D Langridge and J T Morgan — International
        Cryogenic Materials Conference, July 1978, Munich, Germany.
4.   D Evans, J U D Langridge and J T Morgan — International
        Cryogenic Materials Conference, August 1979, Wisconsin, USA.
5.   R Drake and A Siebert — SAMPE Quarterly, July 1975.
6.   'New epoxies for the tough jobs' — Modern Plastics International,
        September 1975.

DYNAMIC MECHANICAL PROPERTIES OF POLY(METHACRYLATES)

AT LOW TEMPERATURES

J Heijboer and M Pineri

TNO and Centre d'Etudes Nucleaires

PO Box 217,Delft,The Netherlands and PO Box 85X,Grenoble

1. INTRODUCTION

The application of plastics in aerospace and cryogenic technology is becoming more and more important. At TNO we had available
a series of systematically chemically modified poly(methacrylates)
for which the dynamic-mechanical properties had been measured down
to liquid nitrogen temperature[1-6]. At CEN, equipment was available
to measure dynamic-mechanical properties from liquid helium temperature upwards[7]. In view of the importance of low temperature
properties we decided to cooperate.
Since molecular mobility might improve the toughness, we searched
for mechanical loss maxima at low temperatures. Furthermore, the
effect of the molecular environment on the location of the n-butyl
loss peak has been studied.

2. EXPERIMENTAL

2.1 PREPARATION OF THE POLYMERS

The polymers were obtained by bulk polymerisation of monomer
between glass plates with azo-bis-isobutyronitrile as initiator[4].
The amount of residual monomer was determined by means of infrared
spectroscopy. Because the effect on dynamic-mechanical properties
below the glass-transition temperature of a few percent of monomer
is only slight, the bulk polymers were not purified.

The specimens were machined from polymer sheets on a lathe;
size of the specimen 25 x 3 x 0.5 mm^3 . Especially for the extremely
brittle poly(cycloalkyl methacrylates) this is a painstaking and
laborious procedure.

2.2 MEASURING EQUIPMENT AND PROCESSING OF DATA

An inverted free oscillating torsional pendulum was used, measurements being carried out during heating at a rate of 60 K/h. A small He-Pressure ($\sim$2.7 KPa) is kept in the pendulum in order to have a good temperature homogeneity during the experiment.

The data are obtained as plots of $\frac{\Delta W}{W}$ and $\frac{G'}{G'_0}$ vs T. W is the stored energy, ΔW is the dissipated energy, G' is the shear modulus and G'_0 is the shear modulus at 4 K. See for an example Figure 1.

The relative loss per cycle is determined from:-

$$\frac{\Delta W}{W} = 1 - \exp\left(- \frac{2}{n} \ln \frac{\Theta_0}{\Theta_n}\right)$$

where Θ_0 and Θ_n are the amplitudes of the oscillations 0 and n.

$\frac{\Delta W}{W}$ is related to the damping factor $Q^{-1} \equiv \tan \delta$ by $\frac{\Delta W}{W} = 1 - \exp(- 2\pi \tan \delta)$.

$\pi \tan \delta = \Lambda$(logarithmic decrement).

The relative shear modulus is obtained from:-

$$\frac{G'}{G'_0} = \frac{P_0^2}{P^2}$$

where P is the period of vibration, P_0 is the same quantity at 4 K, which has been taken as a reference temperature.

Another measure of the losses is the loss modulus in shear G".

$$G'' = G' \tan \delta$$

For small specimens tan δ is more accurate than G", because tan δ does not depend on the dimensions of the specimen.

In general, good agreement was obtained between the previous torsional pendulum data of TNO and the new measurements of CEN; see Figure 2 for an example.

The absolute values of shear moduli of TNO are more accurate than those of CEN, because the size of the TNO specimen (18.0 x 7.5 x 3.5 mm^3) is much larger than that of the CEN specimen (25 x 3 x 0.5 mm^3). In general, no difference larger than 10% is observed between the CEN and TNO values. However, for a very soft specimen, PnHexMA, a difference of about 40% was observed, but this may be due to clamping effects.

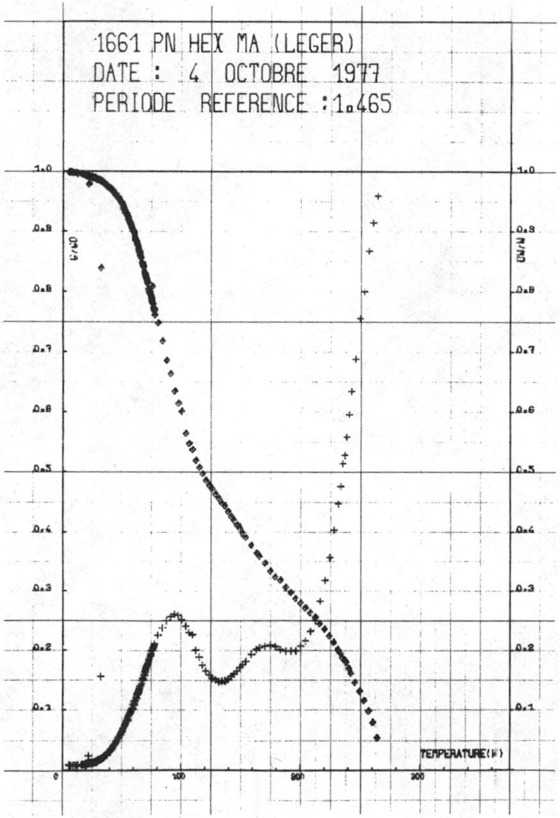

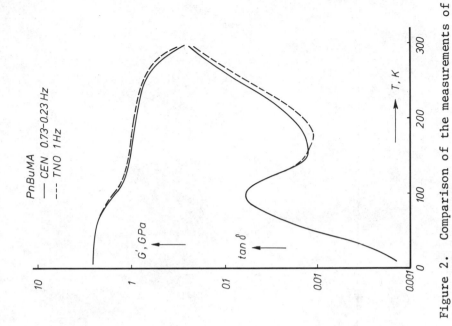

Figure 2. Comparison of the measurements of
CEN and TNO on poly(n-butyl methacrylate).

Figure 1. Example of a computer plot of the
experimental data. Polymer: Poly(n-hexyl methacrylate; "periode reference" is the period
of vibration in s.

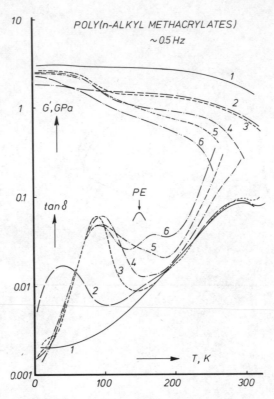

Figure 3.   Shear modulus G' and damping tan δ of poly(n-alkyl methacrylates) as functions of temperature T. PE indicates the location of the loss peak of polyethylene[26].

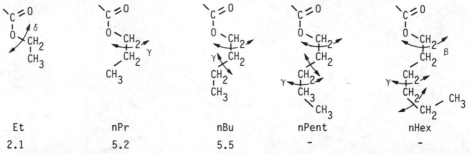

Figure 4.   The n-alkyloxycarbonyl side groups of the poly (methacrylates). The rotations, which probably give rise to the loss maxima, are indicated. Also the activation energies (in kcal/mol), obtained from the frequency dependence of the loss maxima, are given.

In the region of low damping (tan δ < 0.01) the measured damping values are somewhat high in comparison with literature data. This is probably due to internal damping of the equipment.

3. RESULTS AND DISCUSSION

3.1 POLY(n-ALKYL METHACRYLATES)

The general formula of the poly(methacrylates) is given by the formula:-

For the poly(n-alkyl methacrylates) R is $-(CH_2)$ n-1 CH_3. The following members have been measured: n = 1,2,3,4,5 and 6. Only atactic polymers have been investigated. The results are given in Figure 3; both shear modulus G' and damping tan δ are plotted on a logarithmic scale.

Poly(Methyl Methacrylate.), (PMMA), n = 1

Much low temperature data has already been reported[8-13]. It can be seen from Figure 3 that the damping of PMMA decreases gradually with decreasing temperature; only in the region of 10 K is there an indication of a small low temperature maximum. This is in good qualitative agreement with the results of Sinnott[8], Crissman, Sauer and Woodward[9-10] but our damping value is considerably higher (0.002 vs $\sim$0.0003).

In attempting to correlate damping in the region of 10 K with molecular motion, it is necessary to consider the motions of the backbone methyl groups and of the ester methyl groups. Both motions are observed by NMR[12,14,15] and by incoherent inelastic neutron scattering[16,17].

The intramolecular barrier to rotation provides the main contribution to the relaxation process. Assuming quantum mechanical tunnelling, a good agreement between the barriers found by NMR and by neutron scattering is obtained. The barrier for the ester methyl group rotation is only about 1 kcal/mol.

In a detailed investigation, Williams and Eisenberg[13] observed minor mechanical loss maxima between 70 and 170 K and 200-10,000 Hz. In agreement with these data, Waterman[18] found a slight indication of a maximum in the region of 140 K, 10,000 Hz for actactic PMMA.

The pronounced maximum given by Tanabe et al[12] ($\sim$230 K; 10 MHz) is probably due to impurities in the PMMA.

By comparison with a deuterated PMMA sample, Williams and Eisenberg[13] provide strong evidence that the minor maxima they observed are due to motions of backbone methyls corresponding to heterotactic and syndiotactic main chain configurations. They also showed that according to tunnelling theory the curve log νmax vs 1/T max levels off at low temperatures; this means that these maxima cannot be observed in the 1 Hz region. This holds "a fort- iori" for the motion of the ester methyl group, since the barrier for this motion is even lower. On the basis of this information, it must be concluded that the levelling off of the damping near 10 K cannot be ascribed to tunnelling of the ester methyl group. (See, however, the paper of G Federle and S Hunklinger at this conference).

Poly(Ethyl Methacrylate), (PEtMA), n = 2

This polymer shows a pronounced so-called δ-maximum near 40 K, see Figure 3. Wada et al[19] have summarized the literature data about this maximum in a log νmax vs 1/T plot, our value deviates 4 K from this line. The activation energy calculated from this line is 2.1 kcal/mol. From molecular potential calculations they find a barrier of 1.9 kcal/mol for the rotation of the ethyl group around the C-O bond. On this basis, together with arguments derived from relaxation strength, it is concluded that the δ-maximum is due to rotation of the ethyl group around the C-O bond.

It is interesting to compare the δ-peak of PEtMA with that of poly(vinyl ethyl ether), which lies[20] at about 100 K, 1 Hz, tan δ = 0.016. Due to the vicinity of the main chain, the barrier to the motion of the ethyl group is much higher in poly (vinyl ethyl ether).

Poly (n-Propyl Methacrylate), (PnPrMA), n = 3

This is the first polymer in the series, where a large maximum, the so-called γ-maximum, appears. This maximum has already been observed by Hoff, Robinson and Willbourn[21] in 1955. Other data are found in references 1, 5, 22 and 23. The mechanical relaxation has been compared with the dielectric relaxation by Hideshima et al[24]. These authors estimate the activation energy to be about 5.5 kcal/ mol. Waterman et al[5] found from measurements over 6 decades in frequency 5.2 kcal/mol.

Of this polymer, no energy contour map is available. It is not established whether the rotation of only the ethyl end group or the rotation of the n-propyl group as a whole is involved, but a comp- arison with the next member of the series indicates that the main contribution comes from the rotation of the ethyl end group.

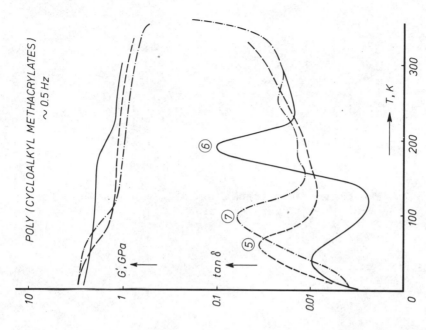

Figure 6. Shear modulus G' and damping tan δ for PCPMA, PCHMA and PCHpMA as functions of temperature T.

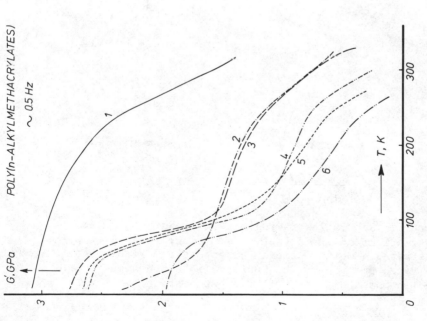

Figure 5. Shear moduli G' of poly(n-alkyl methacrylates), plotted on a linear scale, as functions of temperature.

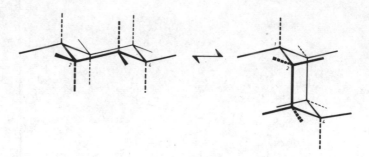

Figure 7. The chair-chair transition of the six-membered ring with carbon atom 1 in a fixed position.

5-ring

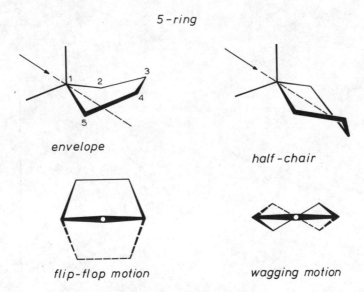

Figure 8. Schematic drawing of "envelope" and "half chair" conformations of the five-membered ring. The lower part illustrates a motion of each conformation, with carbon atom 1 in a fixed position. The conformations are seen in the direction of the arrow in the upper part.

Poly (n-Butyl Methacrylate), (PnBuMA), n = 4

It is seen from Figure 3 that in comparison with PnPrMA, the
γ-peak has become broader and is shifted to a slightly higher
temperature. Again this peak has already been observed by Hoff,
Robinson and Willbourn[21]. The activation energy, found from the
frequency shift, is 5.4 kcal/mol[25]. Wada et al[19] have calculated
an energy contour map for the motion of the n-butyl group. They
come to the conclusion that the motion responsible for the peak is
the rotation of the n-propyl group accompanied by the rotation of
the end ethyl group. The rotation of the whole n-butyl group around
the O-C bond is prevented by steric hindrance of the main chain.

It is difficult to compare this peak with the n-butyl side-group
motions of poly(n-hexene-1), which lies at 120 K[7]. The latter peak
is probably due to a combined motion of the main chain and the side
group.

Poly(n-Pentyl Methacrylate), (PnPentMA), n = 5

To our knowledge, no low temperature measurements on this
polymer have been published. In comparison with PnBuMA the peak is
slightly lower but considerably broader towards higher temperatures.
The low temperature side of this peak is probably due to motion of the
ethyl end group. The increased damping in the region of 160 K might be
ascribed to motion of a larger group, the n-butyl end group of the
n-pentyl group.

Poly (n-Hexyl Methacrylate)*, (PnHexMA), n = 6

In comparison with PnPentMA, the γ -peak has sharpened and a
new peak which we call β, has appeared. The latter peak might be
due to the motion of a n-pentyl group. The sharpening of the γ -peak
and its slight shift to lower temperatures can be explained by
supposing that the contribution of the motion of groups larger than
ethyl becomes smaller.

In Figure 3 the location of the γ-transition of polyethylene
is also given[26]. It is surprising that the latter peak lies at a
slightly lower temperature than the β-peak of PnHexMA, since one
would expect that the motion of a chain part fixed at both ends
would be more difficult than that of a chain end.

Survey of the Poly (n-Alkyl Methacrylates)

Figure 3 shows that pronounced low temperature maxima are
present from n = 2. The low temperature side of the peaks from
n = 3 to n = 6 are very similar; it is likely that in this temperature region damping is caused by motion of the ethyl end group,
rotating around the C-C bond which connects the ethyl end group

---

*Sample kindly provided by Rohm and Haas, GmbH, Darmstadt.

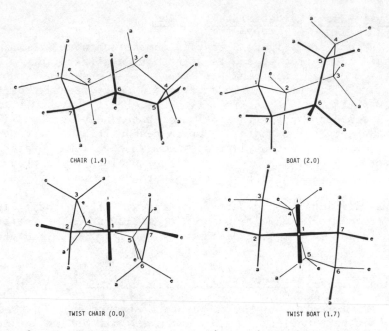

Figure 9. Cycloheptane conformations with equatorial (e) and axial (a) substituents. Value between brackets is the calculated energy in kcal/mol relative to the twist chair conformation.

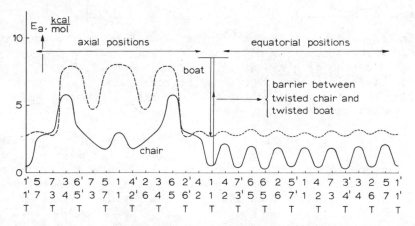

Figure 10. Conformational energies of methylcycloheptane for the chair (—) and the boat (---) pseudo-rotation itinerary according to data given by Hendrickson[37]. The position of the methyl group is indicated below the abscissa: First line: chair conformations; second line: boat conformations; T = twisted conformations.

to the rest of the n-alkyl group. This is indicated in Figure 4. Polymers with n = 3 to n = 6 have such an ethyl group in common. In contrast, the ethyl group for n = 2 is connected to an oxygen atom and this results in a clearly different loss peak, the barrier to rotation around the C-O bond being much smaller.

By comparing with poly (vinyl ethyl ether) it is clear, however, that not only the C-O bond, but also the rest of the local environment influences the potential barrier.

From n = 5, a separate maximum arises, probably due to motion of a group larger than ethyl, but the molecular mechanism of this maximum is not clear.

Table 1 summarizes some data of the loss peaks. The difference between the inverse of the two temperatures where the damping has half its maximum value is taken as a measure of the width of the peak.

The γ-peak broadens considerably from n = 3 to n = 5. This broadening is not continued for n = 6 because of the splitting off of the β-maximum.

Table 1. Low temperature loss maxima of poly (n-alkyl methacrylates)

	TEMPERATURE K	FREQUENCY Hz	HEIGHT tan δ	HALF WIDTH Δ 1000/T K^{-1}
methyl	10?	0.82	0.002?	–
ethyl	39	0.65	0.017	87
n-propyl	90	0.66	0.058	4.6
n-butyl	97	0.56	0.061	5.9
n-pentyl	95	0.65	0.048	8.1
n-hexyl	93	0.56	0.047	7.4
	170	0.41	0.037	–

The moduli are plotted on a linear scale in Figure 5 to show their decrease in more detail. The values for PnHexMA are probably too low; they are about 40% lower than those found with the TNO pendulum.

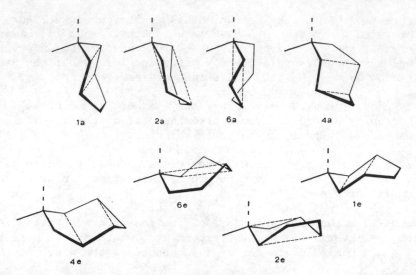

Figure 11. Successive chair conformations of a pseudo-rotation
itinerary of the cycloheptyl ring with one carbon atom in a
fixed position. Positions from 1a to 1e.

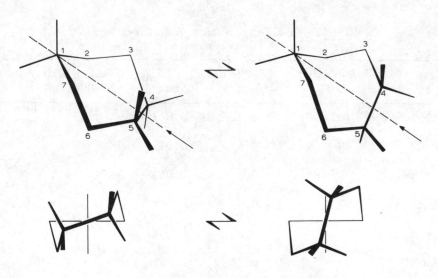

Twist-chair ⟷ Twist boat

Figure 12. Schematic drawing of the twist-chair twist-boat
transition of the seven-membered ring. The direction of the
bonds for the substituted atoms is indicated only for carbon
atoms 1, 4 and 5. The lower part shows the ring views in the
direction of the two fold axis, indicated by an arrow in the
upper part.

The strong decrease in moduli in the region of the loss peaks is evident. Poly (ethyl methacrylate) is the only polymer which shows a strong relaxation below 50 K.

3.2 POLY (CYCLOALKYL METHACRYLATES)

From previous measurements[4] it is evident that poly (cyclopentyl methacrylate) and poly (cyclohexyl methacrylate) have loss peaks below 100 K at 1 Hz. By using higher frequencies, these peaks may in part be observed above 100 K. However, for a better description of these loss peaks, measurements below 100 K are required.

It is known from NMR[27-32] and from computer calculations[33-39] that several rings of intermediate size have a high internal mobility at very low temperature. It seemed worthwhile to investigate whether this mobility results in mechanical loss peaks. Mechanical loss peaks originating from main chain motions can contibute to the impact strength of polymers[40]; it might be an interesting approach to build flexible rings into the main polymer chain to obtain flexibility and good impact strength at low temperature[41].

The formula of the ring compounds investigated is given by the previous formula I of the poly (methacrylates) with

$$R = CH\ (CH_2)_{m-1}$$

The following members have been measured: m = 5, 6, 7, 8, 10 and 12. From m = 6 the polymers are very brittle. Figure 6 shows the results of the compounds with m = 6, 5 and 7. We will discuss the results of each ring-compound separately.

Poly (Cyclohexyl Methacrylate), (PCHMA), m = 6

This polymer shows two maxima, the γ-transition at about 190 K and the δ-transition at about 40 K. These maxima have also been observed by Hoff, Robinson and Willbourn[21] and by Frosini, Maganini, Butta and Baccaredda[42]. The γ-transition is caused by the chair-chair transition of the cyclohexyl ring, which is illustrated in Figure 7[4,43]. The value of the activation energy found from the frequency dependence of the mechanical loss peak over 10 decades, 11.3 kcal/mol, agrees very well with the activation energy of 11.4 kcal/mol, found from NMR measurements for the chair-chair transition of cyclohexane[44].

The molecular origin of the δ-maximum is less clear. One of the authors has previously suggested[4] that it is a rotary-oscillation of the ring as a whole around the C-O bond.

Poly (Cyclopentyl Methacrylate), (PCPMA), m = 5

This polymer shows a pronounced loss peak near 60 K. Considering the low temperature, the relaxation is extremely intense. It has been measured before on a TNO specimen by Roe and Simha[45]. Our data (60 K, tan δ = 0.069) correspond well with theirs (61 K, tan δ = 0.064). We found from the frequency shift an activation energy E_a of 5.1 kcal/mol, whereas from the formula[6]: E_a = (0.060 - 0.046 log ν)T_m, 3.7 kcal/mol is calculated.

To discuss the possible molecular mechanism of the loss peak we have to consider the different molecular conformations of the ring.

The planar conformation of the ring is not favoured because of eclipsed hydrogens. Its energy is about 6 kcal/mol in excess over the envelope - and the half-chair - form (see Figure 8). Pseudo-rotation (exchange of the atom that is not in the plane of the four other atoms of the ring) is very easy, the barrier is at most a few tenths of a kcal[35].

Direct flip-flop motion, sketched in the lower left of Figure 8, is rather unlikely to take place, it would require 6 kcal/mol. The wagging motion sketched in the lower right of Figure 8 is more related to the pseudo-rotational motion and could be the mechanism responsible for the low temperature peak. However, also a (partial) rotation of the whole ring around the O–C_1 bond cannot be excluded.

Poly (Cycloheptyl Methacrylate), (PCHpMA), m = 7

This polymer has a rather broad major loss peak near 100 K. The activation energy, found from its frequency shift, is 6.2 kcal/mol[4].

Like the six-membered ring, the seven-membered ring also has two important conformations, chair and boat. Their twisted forms represent the energy minima[37]; see Figure 9. Both conformations are very flexible and can easily pseudo-rotate.

Figure 10 shows the energy barriers for the pseudo-rotation of methylcycloheptane. The chair conformation has its highest barrier at 6a; from 4a to 2e, the barriers are very low. In Figure 11, the change in position of the ring during such a pseudo-rotation is shown; the change between 4a and 4e is accompanied by a considerable mass transfer. Nevertheless these easy pseudo-rotations do not manifest themselves in mechanical losses: at very low temperatures no considerable damping is observed.

Concerning the molecular mechanism of the loss peak near 100K:

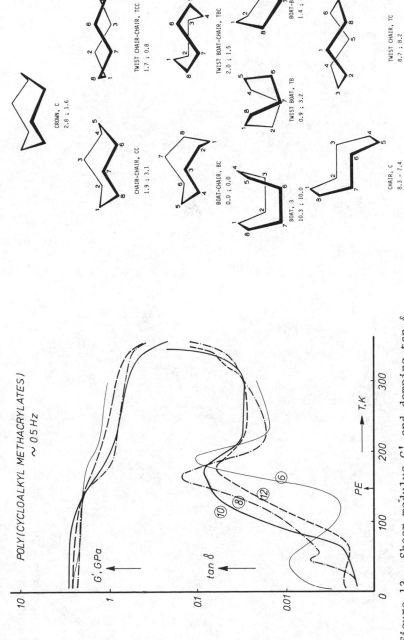

Figure 14. Typical conformations of the cyclo-octane ring. The calculated conformational energies, relative to the BC conformation are given in kcal/mol; first no: data of Hendrickson[37] second no: data of Weigert and Middleton.[30]

Figure 13. Shear modulus G' and damping tan δ for even-membered poly(cycloalkyl methacrylates) as a function of temperature. The number of the carbon atoms in the ring is indicated. The arrow marked PE denotes the temperature of the γ-transition of polyethylene.

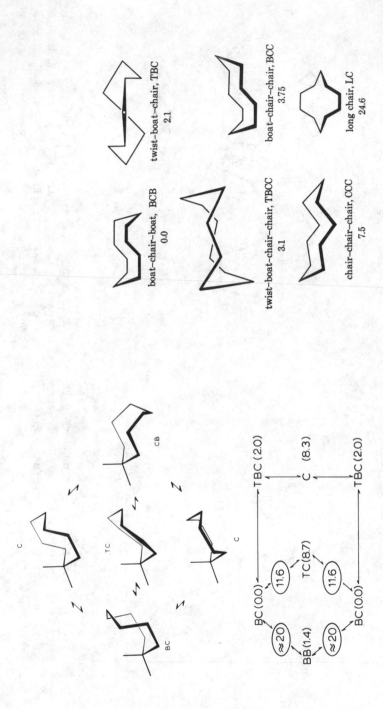

Figure 15. Transition paths of a conformation of one boat-chair family to a conformation of the other boat-chair family. Upper part: schematic drawing of the transitions:

BC → C ⇌ BC and BC ⇌ TC → BC

Lower part: scheme with barriers in kcal/mol according to Hendrickson[37]

Figure 16. Conformations of cyclodecane according to Noe and Roberts[32]. Strain energies, in kcal/mol, relative to the BCB form.

if it is pseudo-rotation of the chair conformation, the peak will
originate from a transition between chair conformations 4a and 2a
(or 5a and 7a), because only this transition has the requisite
energy barrier. The alternative possibility, transition of a chair
to a boat conformation is most likely to take place between twisted
conformations and is sketched in Figure 12. According to
Hendrickson[37] the barrier to this transition is 8 kcal/mol.

The foregoing illustrates how difficult is the exact assignment
of a molecular motion to a damping peak, and moreover that mobility
of a ring is no guarantee of a loss maximum at very low temperatures.

Poly (Cyclo-Octyl Methacrylate), (PCOcMA), m = 8

Figure 13 shows the modulus and damping of the compounds with
m = 6, 8, 10 and 12. The eight-membered ring has the highest loss
peak, it being somewhat broader than that of the six-membered ring,
but much sharper than that of the twelve, ten and seven-membered
rings. It follows, that the molecular mechanism is better defined
for the 8, than for the 7, 10 and 12-membered rings.

The activation energy calculated from the shift of the loss
peak with frequency is 10.6 kcal/mol. The small loss peak near
50 K is due to nitrogen[7].

A large number of conformations is possible for cyclo-octane
[28,29,37]. They are sketched in Figure 14.
Two groups of conformations are important: (a) the boat-chair family,
and (b) the crown family, to which also belong the chair-chair and
twisted chair-chair conformations. The boat-chair family itself
consists of two groups of conformations which easily rotate within
one group, but between the groups there is an energy barrier of
about 8 kcal/mol.

One of the authors has previously suggested[4] that the mechanical
loss peak is probably connected with the transition of one boat-chair
group to the other (boat-chair inversion). Different paths for this
transition are illustrated in Figure 15. The activation energy
calculated from the shift of the mechanical loss peak with frequency
is intermediate between the values calculated for the routes via the
chair and the twisted chair conformation. Anet[29] considers a value
of 10.6 kcal/mol to be too high to be connected to a boat-chair ring
inversion and considers a boat-chair to twist chair-chair inter-
conversion as a more likely process. The barrier for this inter-
conversion is, according to Hendrickson, 11.4 kcal/mol.

We conclude that for this ring also, pseudo-rotation does not
result in a loss peak at very low temperatures (<50 K).

Poly (Cyclodecyl Methacrylate), (PCDecMA), m = 10

This polymer has a very broad loss peak; it is composed of several loss peaks. Figure 16 shows conformations of cyclodecane, together with their strain energies, according to Noe and Roberts[32]. Anet, Cheng and Wagner[47] conclude on the basis of NMR data that the conformational free energy barrier of cyclodecane is approximately 6 kcal/mol.

The estimated activation energy for the mechanical loss peak is about 10 kcal/mol.
On the basis of the mechanical data we can only conclude that the molecular mechanism is complex and more than one mechanism is active.

Poly (Cyclododecyl Methacrylate), (PCDodMA), m = 12

This polymer has a broad loss peak in the region of 170K, but it is lower than the peaks for the other even-membered rings. This indicates a lower mobility. Indeed the steric hindrance in the inner part of the ring is rather large[39]. Anet and Rawdah[48] conclude on the basis of calculations that the [3333] conformation is preferred. They also indicate a pseudo-rotation itinerary for this conformation, with a barrier of about 8 kcal (see Figure 17). This is in agreement with the barrier they obtain from NMR data.
From the frequency shift of the loss peak we calculate an activation energy of 10.0 kcal/mol, which is appreciably higher.

The 12-membered ring also shows a minor loss peak in the region of 15 K.

A survey of the results on the poly (cycloalkyl methacrylates) is given in Table 2. The activation energy E_a has been calculated from the shift of the loss peak with frequency.

Cowie and McEven[49] have measured poly (itaconic acid) esters containing pendant cycloalkyl groups by means of a torsional braid analyser near 1 Hz. They observe maxima for the 5-, 6-, 7-, 8- and 12-membered rings respectively at 125, 186, 113, 175 and 179 K. With the exception of the 5-membered ring this sequence is in qualitative agreement with our data.

It is seen from Figure 13 that the loss peaks of the 8-, 10-, and 12-membered rings overlap the temperature region of the γ-peak of polyethylene; this means that the mobility of these rings does not differ very strongly from that of the polyethylene chain. The seven and five-membered rings are more mobile.
It is seen from Table 2 that the cyclohexyl ring shows the narrowest loss peak, followed by the cyclo-octyl ring. Only the molecular motion corresponding to the cyclohexyl loss peak is well defined and definitely established; for the other loss peaks the exact mechanism

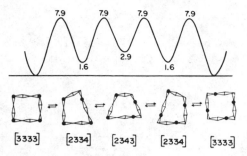

Figure 17. Pseudo-rotation itinerary for the {3333} conformation of cyclododecane according to Anet and Rawdah[48]. Intermediate conformations, but not transition states, are shown. Black circles are carbon labels.

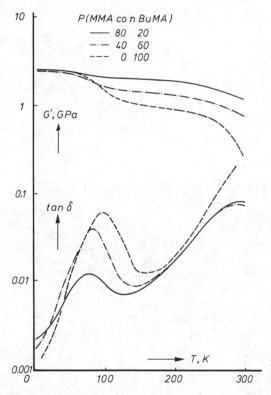

Figure 18. Shear modulus G' and damping tan δ of two copolymers of n-butyl methacrylate and methyl methacrylate as functions of temperature T. The composition is indicated in weight percent. For comparison the curves of PnBuMA are also given.

of the corresponding molecular motion is still a matter of guesswork.

In spite of low barriers for pseudo-rotation, no corresponding pronounced loss peaks at low temperatures are observed.

Table 2. Low temperature loss maxima of poly(cycloalkyl methacrylates) at about 1 Hz.

	TEMP. K	FREQUENCY Hz	HEIGHT tan δ	E_a kcal/mol	HALF WIDTH Δ 1000/T,K^{-1}
cyclopentyl	61	0.64	0.064	5.1	18.0
cyclohexyl γ	188	0.64	0.108	11.3	0.91
δ	41	0.77	0.0096	2.1	112
cycloheptyl	97	0.61	0.062	6.2	6.5
cyclo-octyl	165	0.59	0.132	10.6	1.4
	163	0.58	0.150		1.3
cyclodecyl	170	0.70	0.085	–	3.3
cyclododecyl γ	174	0.90	0.062	10.0	2.2
δ ~	15	–	0.002	–	–

3.3 THE EFFECT OF THE MOLECULAR ENVIRONMENT ON THE n-BUTYL MAXIMUM

Kolarik[50] observed an increase in the temperature of the γ-loss maximum in the copolymers of methyl methacrylate and n-butyl methacrylate with increasing content of nBuMA. He suggests that this result indicates that free volume is an important factor for the temperature location of secondary loss maxima.
Free volume decreases during physical ageing. Struik[51], however, found practically no effect of ageing on the location of the secondary loss peaks. One of us[4] showed that the location of the γ-loss maximum of the cyclohexyl group is practically independent of the molecular environment. We were therefore interested to examine more closely the effect of the molecular environment, particularly free volume, on the location of the maximum caused by the motion of the n-butyl group.

Figure 18 shows that we too find for the copolymer series MMA-nBuMA a shift to higher temperature with increasing nBuMA content.

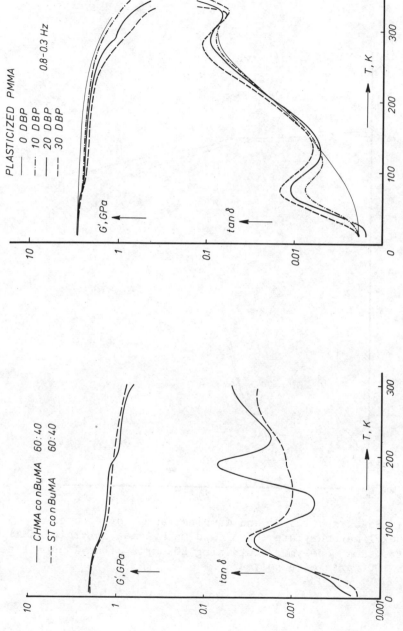

Figure 20. Shear modulus and damping of poly-
(methyl methacrylate) plasticised with different
amounts of dibutyl phthalate (DBP); parts DBP per
100 parts PMMA.

Figure 19. Shear modulus G' and damping tan δ
of copolymers of n-butyl methacrylate with cyclohexyl methacrylate (CHMA) and styrene (ST). Composition in weight percent.

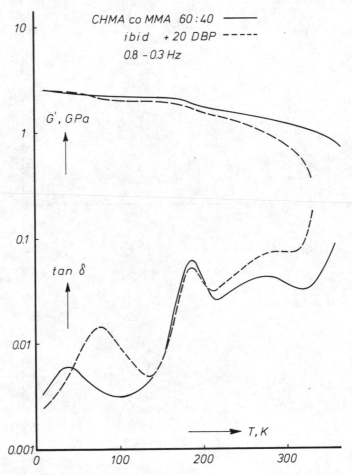

Figure 21. Shear modulus G' and damping tan δ for a copolymer of cyclohexyl methacrylate (CHMA) and methyl methacrylate (MMA) and for the same copolymer containing 20 parts dibutyl phthalate (DBP) per 100 parts of copolymer.

Figure 19 shows, however, that the γ-maximum in a copolymer with CHMA lies at about the same temperature as that of a copolymer with styrene (it is interesting that in the copolymer with CHMA the δ-maximum of CHMA has disappeared).

Figure 20 shows that also polymers, in which the n-butyl group is present in the plasticiser molecule instead of in the polymer molecule, have a n-butyl maximum.

Figure 21 shows the effect of the DBP on a CHMA-MMA copolymer. Again the maximum of the n-butyl group of the plasticiser is manifest and the δ-maximum of the cyclohexyl group has disappeared. It is also seen that the location of the γ-cyclohexyl maximum is not affected by the plasticiser.

Table 3 gives a survey of the data of the polymers mentioned so far, together with those of a copolymer with butyl acrylate (BuA).

In Figure 22, the temperature of the maxima of G" vs T are plotted as a function of the n-butyl concentration. We find the same effect as Kolarik for the n-butyl maximum, viz. an increase of the temperature of the loss maximum with increasing concentration of the n-butyl groups. However, since the composition of our polymers varies much more widely than that of Kolarik's, it is very unlikely that the increase in temperature of the loss maximum can be described to a decrease in free volume. It is, e.g., known that a small amount of DBP in PMMA causes volume contraction; nevertheless the points for the polymers containing DBP fall on the same line. The concentration of the n-butyl groups is clearly the factor which governs the temperature of the loss maximum.
The increase in the temperature with increasing n-butyl concentration might be due to mutual interaction of the n-butyl groups.

Figure 22 also shows that, in contrast, the concentration of the cyclohexyl group has no effect on the temperature of the cyclohexyl loss maximum. The cyclohexyl groups do not seem to interact with each other.

Figure 23 gives the height of the n-butyl loss maximum (expressed as the loss modulus G''_m) as a function of the n-butyl concentration. From this figure it follows that a n-butyl group present in a plasticiser molecule gives the same contribution to the mechanical losses as the n-butyl group present in the polymer molecule.

Similarly, the relaxation strength of the cyclohexyl group is the same whether the group is present in the polymer or in a plasticiser molecule[4]. This indicates that for mechanical processes in glassy polymers, the interaction between the stress and the

Table 3.　Loss peaks of the n-butyl group

Composition	Density(23°) kg/m³	{Bu} mol/l	tan δ-maximum		G" maximum	
			T_m K	height 1000 tan δ	T_m K	G''_m MPa
100 PnBuMA	1056	7.42	97	61	90	98
100 PMMA 10 DBP	1183	0.85	74	7.6	74	20
100 PMMA 20 DBP	1177	1.41	76	11	76	28
100 PMMA 30 DBP	1171	1.94	76	15	74	39
60 CHMA co 40 MMA + 20 DBP	1133	1.36	76	14.5	74	32
60 CHMA co 40 n Bu MA	1088	3.05	81	28	78	42(48)*
80 MMA co 20 n Bu MA	1161	1.64	77	12	76	28
40 MMA co 60 n Bu MA	1105	4.66	83	38	82	69
60 St co 40 n Bu MA	1065	3.00	81	34	79	52
70 MMA co 30 n Bu A	1148	2.69	80	24	78	45

* Modulus value corrected.
Composition in weight percent; amount of plasticiser per 100 polymer.

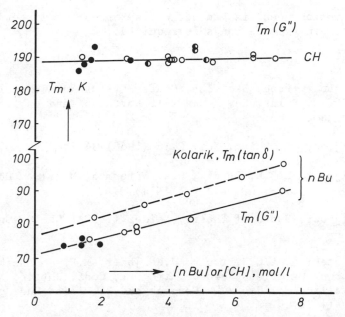

Figure 22. Temperatures Tm of the maxima of the loss modulus of
copolymers of containing cyclohexyl (CH) and n-butyl (nBu)
groups as functions of the concentration of the groups.
For comparison the maxima of tan δ in a copolymer series
nBuMA-MMA are given[50].

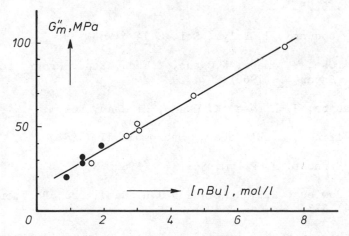

Figure 23. Height G_m'' of the n-butyl loss maximum as a function
of the n-butyl concentration for polymers containing the n-butyl
group in a polymer molecule (open circles) or in a plasticiser
molecule (filled circles).

molecular motion goes via Van der Waals forces; no direct interaction via main valence bonds is required.

REFERENCES

1. J.Heijboer, Proc. Sec. Int. Conf. Physics Non-Crystalline
    Solids, W. Prins, Ed., page 231, North Holland Publishing Co.,
    Amsterdam, 1965.

2. J.Heijboer, J. Polym. Sci. C 16 (1968) 3413.

3. J.Heijboer, L.C.E. Struik, H.A. Waterman, M.P. van Duijkeren,
    J. Macromol. Sci. Phys. B 5 (1971) 375.

4. J.Heijboer, Doctoral Thesis, Leiden (1972), CL-TNO Communication
    435.

5. H.A.Waterman, L.C.E.Struik, J.Heijboer, M.P.van Duijkeren,
    Amorphous Materials, Proc. Third Int. Conf. Physics Non-
    Crystalline Solids, R.W. Douglas and B. Ellis, Eds, page 29,
    Wiley (1972).

6. J. Heijboer, Intern. J. Polym. Mat. 6 (1977) 11.

7. M. Pineri, Polymer 16 (1975) 595.

8. K.M. Sinnott, J. Polym. Sci. 35 (1959) 273.

9. J.M. Crissman, J.A. Sauer, A.E. Woodward, J. Polym. Sci. A 2
    (1964) 5075.

10. A.E. Woodward, J. Polym. Sci. C 14 (1966) 89.

11. J. Hirose, Y.Tanabe, K.Okano, Y.Wada, Rept. Progr. Polym.
    Phys. Japan 11 (1968) 267.

12. Y. Tanabe, J. Hirose, K. Okano, Y. Wada, Polym. J. 1 (1970) 107.

13. J. Williams, A. Eisenberg, Macromol. 11 (1978) 700.

14. K.M. Sinnott, J. Polym. Sci. 42 (1960) 3.

15. J.G. Powless, B.I. Hunt and D.J.H. Sandiford, Polymer 5
    (1964) 323.

16. J.S. Higgings, G.Allen and P. Brier, Polymer 13 (1972) 157.

17. G. Allen, C.J. Wright and J.S. Higgins, Polymer 15 (1974) 319.

18. H.A. Waterman, CL-TNO, personal communication.

19.   K. Shimuzu, O. Yana, Y. Wada and Y. Kawamura, J. Polym. Sci., Polym. Phys. 11 (1973) 1641.

20.   R.G. Saba, J.A. Sauer and A.E. Woodward, J. Polym. Sci., Polym. Phys. 1 (1963) 1483.

21.   E.A.W. Hoff, D.W. Robinson, and A.H. Willbourn, J. Polym. Sci. 18 (1955) 161.

22.   T.Hideshima, Rep. Progr. Polym. Phys. Japan 6 (1963) 143.

23.   F. Lednicky and J. Janacek, J. Macromol. Sci.-Phys. B 5 (1971) 335.

24.   M. Kakizaki, K. Aoyama and T. Hideshima, Rep. Progr. Polym. Phys. Japan 17 (1974) 379.

25.   J. Heijboer, Ann. N.Y. Acad. Sci. 279 (1976) 104.

26.   M. Pineri, P. Berticat and E. Marchal, J. Polym. Sci., Polym. Phys. 14 (1976) 1325.

27.   E.S. Glazer,R. Knorr, C. Gauter, J.D. Roberts, J. Am. Chem. Soc. 94 (1972) 6026.

28.   F.A.L. Anet and V.J. Basus, J. Am. Chem. Soc. 95 (1973) 4424.

29.   F.A.L. Anet, Top. Current Chem. 45 (1974) 169.

30.   F.J. Weigert and W.J. Middleton, J. Org. Chem. 45 (1980) 3289.

31.   F.A.L. Anet and J.J. Wagner, J.Am. Chem. Soc. 93 (1971) 5266.

32.   E.A. Noe and J.D. Roberts, J. Am. Chem. Soc. 94 (1972) 2020.

33.   E.L. Eliel, N.L. Allinger, S.J. Angyal and G.A. Morrison, Conformational Analysis, Interscience, New York, 1965.

34.   N.L. Allinger, J.A. Hirsch, M.A. Miller, I.J. Timinski and F.A. Van-Catledge, J. Am. Chem. Soc. 90 (1968) 1199.

35.   J.B. Hendrickson, J. Am. Chem. Soc. 83 (1961) 4537.

36.   J.B. Hendrickson, J. Am. Chem. Soc. 86 (1964) 4854.

37.   J.B. Hendrickson, J. Am. Chem. Soc. 89 (1967) 7036,7043 and 7047.

38.   H.M. Pickett and H.L. Strauss, J. Am. Chem. Soc. 92 (1970) 7281.

39. J. Dale, Acta Chem. Scand. 27 (1973) 1115 and 1130.

40. J. Heijboer, J. Polym. Sci. C 16 (1968) 3755.

41. N. Matsumoto, H. Daimon and J. Kumanotani, J. Polym. Sci.,
     Polym. Chem. 18 (1980) 1665.

42. V. Frosini, P. Maganini, E. Butta and M. Baccaredda, Kolloid-Z.
     213 (1966) 115.

43. J. Heijboer, in Molecular basis of transitions and relaxations,
     Ed. by D. J. Meier, Gordon and Breach, London, 1978, page 297
     (Midland Macromol. Monographs, Vol. 4).

44. G. Binsch, Top. Stereochem. 3 (1968) 67.

45. J.H. Roe and R. Simha, Intern. J. Polym. Mat. 3 (1974) 193.

46. R.L. Hilderbrandt, J.D. Wieser and L.K. Montgomery, J. Am.
     Chem. Soc. 95 (1973) 8589.

47. F.A.L. Anet, A.K. Cheng and J.J. Wagner, J. Am. Chem. Soc. 94
     (1972) 9250.

48. F.A.L. Anet and T.N. Rawdah, J. Am. Chem. Soc. 100 (1978) 7166.

49. J.M.G. Cowie and I.J. McEwen, Europhys. Conf. Abstr. 4 A (1980)
     123. Europhysics Conference on Macromolecular Physics,
     Noordwijkerhout, 21-25 April 1980.

50. J. Kolarik, J. Macromol. Sci.-Phys. B 5 (1971) 355.

51. L.C.E. Struik, Physical aging in amorphous polymers and other
     materials, Elsevier, Amsterdam, 1978.

DIELECTRIC LOSS DUE TO ANTIOXIDANTS IN POLYOLEFINS

R Isnard, G Frossati, J Gilchrist and H Godfrin

Centre de Recherches sur les Très Basses Températures
C.N.R.S., BP 166 X
38042 Grenoble-Cedex, France

INTRODUCTION

Thomas and King[1] reported in 1975 that the presence in polyethylene of certain antioxidants of the substituted phenol type
caused a dielectric relaxation at 4.2 K. Although peak loss factor
occurred at several kHz, the loss enhancement at 50-60 Hz was
significant in view of the possible use of polyethylene tapes for
the insulation of a superconducting power transmission line. It
has now been found that all antioxidants that are derivatives of 2,6
di-tert-butyl phenol ("DBP" - see figure 1) cause such a relaxation,
but that any other phenol or non-phenol antioxidant may be used
without causing significant loss enhancement at cryogenic temperatures.[2] DBP and its derivatives cause similar relaxations when
added to polyethylene, polypropylene or other polyolefins, or
dissolved in paraffin or decalin.[3] By an appropriate isotopic
substitution it has been demonstrated that the effect is due to the
hindered rotation of the hydroxyl groups of the molecules,[4] so in
fact it is now clear that Thomas and King rediscovered at cryogenic
temperatures a dielectric relaxation which had been known since
1957. Davies and Meakins[5] then reported that a solution of 2, 4,
6 tri-tert-butyl phenol in decalin had two resolved relaxations,
and were able to identify the faster one (30 GHz loss peak at
ambient temperature) with hydroxyl group rotations and the slower
one with rotations of entire molecules. There are strong grounds
for supposing that at cryogenic temperatures and in solid matrices,
rotation of entire molecules will be negligible, but that the
hydroxyl groups will continue to rotate by quantum mechanical
tunneling. The tunneling hypothesis was made by Thomas and King[1]
and was confirmed by the deuteration experiments[3,4] when it was
found that a DBP derivative with a deuteroxy group relaxed 10^4

117

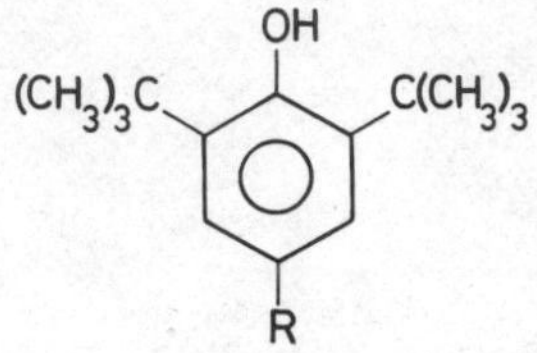

Figure 1

times more slowly than a DBP derivative with a (proton) hydroxy group - the biggest isotope shift of a dielectric relaxation that we know.

Figure 1 shows the structural formula of 2,6 di-tert-butyl phenol ("DBP") and its antioxidant derivatives. DBP is the molecule illustrated with R an H atom. The simplest commercial antioxidant, BHT, has R a methyl group. The molecule studied by Davies and Meakins[5] had R a tert-butyl group. Other antioxidants have R a long chain (example: Irganox 1076) or else a methylene or other group which serves to join two or more DBP molecules together "tail to tail" (examples: Ionox 220, Ionox 330 or ethyl 330, Irganox 1010).

Studies of a wide variety of phenol derivatives diluted in various matrices or solvents[3] have revealed that the relaxations are never so well developed as in the case of DBP and its derivatives in polyolefins. It seems that these are the best systems that can be devised to study this remarkable physical phenomenon.

Our present purpose is firstly to restate briefly the basic finding of electrical engineering interest, and in doing so to set the record straight in view of some erroneous conclusions in the recent literature; secondly to suggest how the effects may be used in sub-Kelvin thermometry.

ANTIOXIDANTS IN POLYOLEFINS

Polyolefins are generally stabilised against oxidative degradation by the addition of a milli-molar concentration (100-1000 ppm by weight) of an antioxidant which is usually a phenol derivative, but may also be an amine. We recently reported[6] that a low-density polyethylene sample containing an amine antioxidant showed no sign of extra loss, nor did two other such samples containing other anti-oxidants that were neither phenols nor amines. Two samples containing DBP derivatives (Irganox 1010 and Irganox 1035) exhibited the relaxation, but so did three samples containing the non-DBP phenols illustrated in figure 2. It was later discovered however, that this series of samples had been prepared by adding the Nonox WSP etc to a polythylene which already contained an antioxidant, so the conclusions were invalid. New polyethylene samples containing Topanol CA, Nonox WSP and Santonox R were examined and

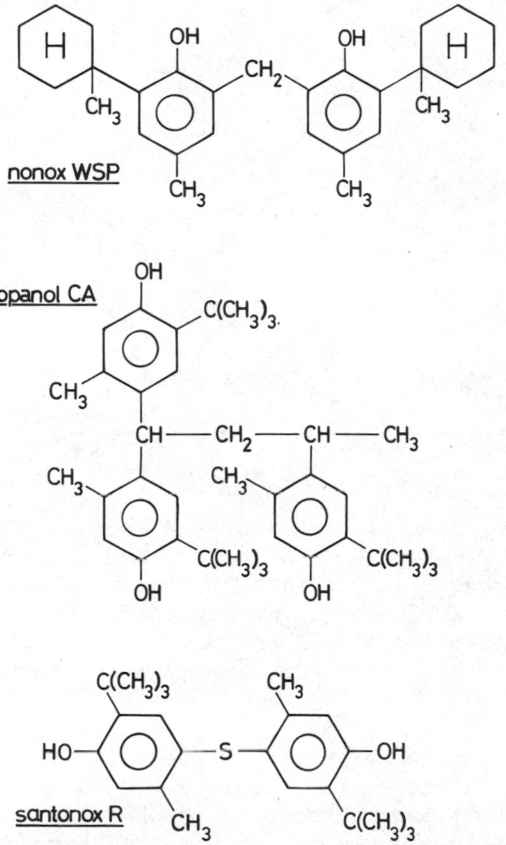

Figure 2

found not to exhibit any detectable relaxation,[2] and it appeared
that the effect was probably an exclusive feature of DBP derivatives.
In addition to the Ethyl 330 (or Ionox 330) originally reported by
Thomas and King,[1] it has now been reliably shown that the following
DBP derivatives all cause the relaxation:[2-4] DBP itself, BHT, 2, 4,
6 tri-tert-butyl phenol (at cryogenic temperatures as well as
ambient), Ionox 220 and Irganox 1076. In the case of solutions in
paraffin or decalin there is a proportionality of approximately
0.09 radians/M between the peak loss angle at 4.2 K and the concentration in g.moles/ℓ, over a fairly wide concentration range. A
series of seven polymer samples also had the same mean value of
this ratio, but individual values ranged from 0.06 to 0.17 radians/M.[4]

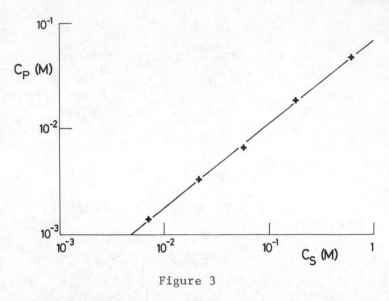

Figure 3

Figure 2 shows the structural formulae of three non-DBP phenol antioxidants, Nonox WSP is 2,2'-methylene-bis(4-methyl-6-(1-methyl-cyclohexyl)phenol); Topanol CA is tris 1,1,3(2-methyl-4-hydroxy-5-tert-butyl)butane; Santonox R is 4,4'-thiobis (6-tert-butyl-meta-cresol).

Figure 3 shows the experimental results from laboratory preparation of polyethylene films containing BHT. The 100 µm thick films were soaked in cyclohexane solutions of BHT for 10 days at $22^{\circ}C$, then washed very rapidly in pure cyclohexane. The resulting molar concentration, C_p in the film (deduced from the 4.2 K loss peak by assuming 0.09 radians/M) is plotted as a function of the molar concentration of the solution. If the washing in pure cyclohexane was prolonged, even for a minute or two at ambient temperature, most of the BHT was lost.

In fact the only negative results with DBP derivatives were the early reports on BHT in polyethylene[1,7] and we have suggested that these were due to the inadvertent loss of the BHT by diffusion and evaporation.[2,4] Moisan[8] recently completed a study of the diffusivity of BHT and other antioxidants in polyethylene and it is clear that BHT can be lost very rapidly if the sample is heated above ambient temperatures at any stage during the preparation. In fact Forsyth et al[9] first positively reported the increased low temperature loss due to BHT in polyethylene (Ionol of reference 9 is BHT under another name). They also first showed the absence of increased loss due to Topanol CA (Topanol in reference 9) and the absence of increased loss due to dilauryl-thiodipropionate ("DLTDP") – a non-phenol additive generally used together with a phenol anti-oxidant. The list of antioxidants suitable for use with a poly-ethylene or polypropylene tape for superconducting cable insulation therefore includes the three illustrated in figure 2, with which DLTDP may be used if required, and amines and other non-phenol antioxidants.

Yano et al[7] also reported the absence of relaxation in the case of Ionox 330 in polystyrene. We have confirmed[3,4] that the effect is suppressed by the presence of a large concentration of aromatic groups, presumably on account of a weak interaction of the hydrogen-bonding type between the DBP and the other benzene rings.

OTHER PHENOLS

For our earlier studies we relied on polymer samples supplied to us, with their antioxidants, by various producers. Later we adopted the technique of introducing DBP etc into polymer films by diffusion. Representative results of this technique are illustrated by figure 3. By the same means we also introduced, for the purpose of study, a variety of other molecules whose addition to polyolefins has no commercial usefulness of which we are aware. Amongst these were various other classes of phenol derivatives. The results could generally be readily understood in terms of the accepted structure and properties of these molecules as deduced from spectroscopic studies. The details are reported elsewhere[3] but figure 4 is illustrative of our findings. Phenol has a slower and much weaker relaxation than DBP, on account, respectively, of its higher barrier to hydroxyl rotation and its much greater tendency to associate by hydrogen bonding, so that only a few molecules will be found, unassociated in the polypropylene matrix. 2, 4, 6 trichlorophenol relaxes strongly but very much more slowly than the others. Its intramolecular bonds. Unsymmetrically substituted phenols (figure 2) sometimes have weak relaxations, but only 2, 6 di-ortho-substituted phenols have strong ones.

POSSIBLE USE IN THERMOMETRY

Capacitance thermometry is an attractive alternative to resistance thermometry where a convenient rapid-response device is required for sub-Kelvin temperatures. Firstly, a high temperature coefficient of capacitance is not necessarily associated with a high loss factor of the dielectric at the measuring frequency, since the variation may result from a temperature-dependent resonance or relaxation occurring at a much higher frequency. By a suitable choice of dielectric material and measuring frequency an appreciable voltage may be used without excessive self-heating of the thermometer. Secondly, capacitance thermometers tend to be much less sensitive to magnetic fields than resistance thermometers. The sensitivities of various types of capacitance thermometers between 1 mK and 100 mK are compared in figure 5. The comparison is somewhat arbitrary because the sensitivities depend on the measuring field strength and frequency as well as the material.

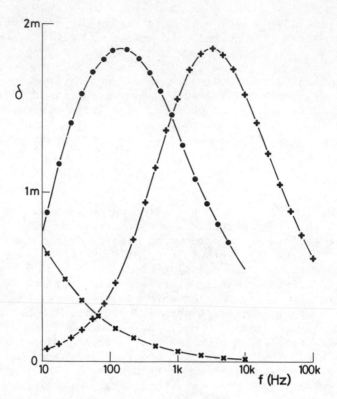

Figure 4

The performance of KCl-OH below 65 mK, to our knowledge, has not
yet been reported but the sensitivity of the BHT-doped polyolefin
compares favourably with the others between 5 mK and 50 mK. The
major disadvantages of doped polymer thermometers are their
dimensional and compositional instability at ambient temperature,
meaning that they require calibration for each run. The corresponding advantages are that they can be easily fabricated and
adapted to the particular thermometry problem (for example glued
or wrapped round a cylindrical sample). The prospects for
capacitance thermometry in the sub-millikelvin range also look
promising. Materials which are likely to be sensitive in this
range include polyolefins containing 2, 6 di-halo-phenol derivatives.
These should relax many orders of magnitude more slowly than the
convenient measuring frequencies of about 1 kHz, so loss factor
should be minimal.

    Figure 4 shows the loss angles in milliradians, at 4.2 K of
diluted phenols: + ... 2, 6 di-tert-butyl phenol at 2×10^{-2} M
concentration in paraffin solution: x ... 2, 4, 6 trichlorophenol

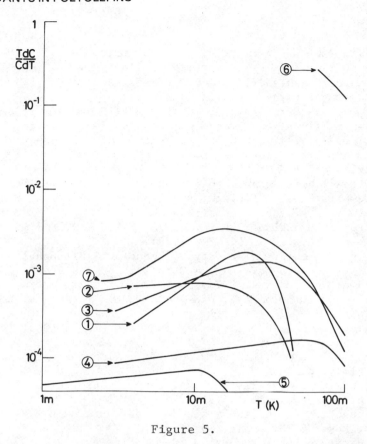

Figure 5.

at 4, 5 x 10^{-2} M concentration in paraffin; • ... phenol in polypropylene at limit of solubility (scale multiplied by 50; the
peak loss was 37 microradians).

Figure 5 shows the sensitivities of various capacitance
thermometer materials: (1) $SrTiO_3$ glass ceramic as reported in
refs. 10 and 11;   (2) borosilicate glass at 110 Hz and 91 mV/170 μm
as in ref. 12;   (3) soda glass at 3.5 kHz and 200 mV/160 μm;   (4)
vitreous silica with 0.044 M hydroxyl concentration at 11 kHz and
≤ 300 mV/500 μm as in ref. 13;   (5) kapton (typical dipolar polymer)
at 1.1 kHz and 91 mV/14 μm;   (6) KCl with 0.004 M hydroxyl concentration as in ref. 14;   (7) poly-4-methyl-1-pentene containing
approximately 0.050 M of BHT at 1.55 kHz and 200 mV/130 μm as in
ref. 3.

REFERENCES

1.   R A Thomas and C N King, Appl. Phys. Lett. 26, 406 (1975).
2.   J le G Gilchrist, Cryogenics 19, 281 (1979).
3.   R Isnard, G Frossati, J le G Gilchrist and H Godfrin, Chemical
          Physics (in the press).
4.   J le G Gilchrist, Dielectric Materials, Measurements and
          Applications (conference publication 177), London, IEE
          (1979) p. 211, and a longer version of this paper to appear
          in IEE Proceedings, Part A.
5.   M Davies and R J Meakins, J. Chem. Phys. 26, 1584 (1957).
6.   J le G Gilchrist, Non metallic Materials and Composites at Low
          Temperatures, A F Clark, R P Reed and G Hartwig, Editors,
          New York, Plenum Press (1979) p. 103.
7.   O Yano, T Kamoshida, S Sekiyama and Y Wada, J. Polym. Sci.
          Polym. Phys. Ed. 16, 679 (1978).
8.   Y Moisan, Thesis, University of Rennes, 1979 (unpublished).
9.   E B Forsyth, A J McNerney, A C Muller and S J Rigby, IEE Trans.
          Power Appl. & Syst. 97, 734 (1978).
10.   D Bakalyar, R Swinehart, W Weyhmann and W N Lawless, Rev. Sci.
          Inst. 43, 1221 (1972).
11.   D Bakalyar, R Swinehart, W Weyhmann and W N Lawless, Low
          Temperature Physics - LT 13, K D Timmerhaus, W J O'Sullivan
          and E F Hammel, Editors, New York, Plenum Press (1974)
          vol. 4, p. 646.
12.   G Frossati, R Maynard, R Rammal and D Thoulouze, J. Physique
          Lettres 38, L-153 (1977).
13.   G Frossati, J le G Gilchrist, J C Lasjaunias and W Meyer,
          J. Phys. C 10, L-515 (1977).
14.   J B Hartmann and T F McNelly, Rev. Sci. Inst. 48, 1072 (1977).

LOW TEMPERATURE FRACTURE MEASUREMENTS ON POLYETHYLENE

IN COMPARISON WITH EPOXIDE RESINS

Bernhard Kneifel

Karlsruhe Nuclear Res. Center
Institut für Technische Physik
Karlsruhe, FRG

INTRODUCTION

The mechanical properties of several epoxide resins (EP), linear high-molecular isotropic polyethylenes (HDPE) and stretched HDPE were determined between 2 and 293 Kelvin. This paper will concentrate on the tensile strength σ_B, the critical stress intensity factor K_{1c} and the energy of fracture γ.

I)   The measurements showed that σ_B is approximately equal for the polymers under consideration. But K_{1c} and γ of HDPE were found to be greater by one order of magnitude than the respective values for EP. This can be seen in connection with the ductile stress-strain behaviour of HDPE even at 4K, whereas EP's fracture without yielding.

II) σ_B shows only a small temperature dependence well below the glass transition, whereas K_{1c} and γ are strongly dependent on temperature even below 4K.

III) More insight into fracture mechanics on HDPE can be gained from studying stretched materials. The lamellar crystallites of HDPE are aligned perpendicular to the stretching direction. A more ductile behaviour in the stretching direction and a great anisotropy of σ_B, K_{1c} and γ were found for HDPE stretched by only 200%.

MATERIALS

Epoxide resins consist of cross-linked and entangled polymer chains forming a three-dimensional amorphous network. Four epoxide resins[1] with differing mean cross-link distances of between 1.5 and 7.5 nm were selected for testing.

Fig 1: a) Fracture surface of stretched HDPE (GUR) (LM).

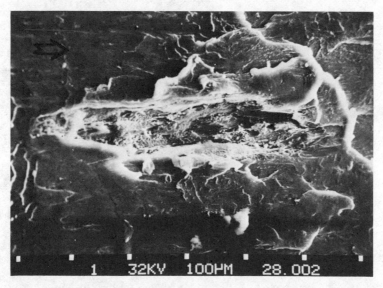

Fig 1: b) SEM-picture of a large defect penetrating the fracture
surface.
(Arrow indicates crack and stretching direction.)

Two types of linear polyethylene were used for measurements, one having a molecular weight of 5 x 10^6, processed by extrusion[2] and the second one having a molecular weight of 20 x 10^6, fabricated by pressure sintering.[3] The crystallinity of both types is approximately 65% by volume.

From the second type of HDPE, aligned samples were produced by stretching the material by 200% when cooling it down from 373K. Due to plastic flow there is some stable alignment of the crystallites and the amorphous chains. In the stretched material one finds under the Scanning Electron Microscope (SEM) that the lamellar layers are mainly oriented perpendicular to the direction of stretching[4] and that there are thin voids of some 10^{-7}m long in the stretching direction. These may be caused by the fabrication process. Under a visual microscope one can see holes of 0.5 to 1 mm in the direction of stretching which are distributed in the whole volume (Fig 1). These were not visible in the isotropic unstretched material.

EXPERIMENTAL

The tensile strength σ_B was measured using cylindrical rod specimens in low-temperature testing machines.[5] The stress-intensity factor K_{1c} was measured with compact tension (CT) specimens (Fig 2) in a very stiff testing machine. This machine,[6] designed for 100 kN, is used only up to 5% of its maximum load. The spindle is driven by an electric motor[12] which ensures a uniform strain velocity. An electrical circuit stops the motor within microseconds; it is triggered by a certain load drop caused by crack propagation.[10] The load and deflection are measured by strain gauges. All machines were driven at a constant strain rate.

For the tensile strength measurements surface effects of the rod samples were minimised by polishing. Many specimens were broken by cracks starting inside the sample, at points where material inhomogeneity or small holes could be seen.

The rod specimens are subjected to a uniform load over the whole cross-section but stress concentrates in an uncontrolled manner at defects like holes, voids and crazes. A crack, once started, cannot be stopped because the elastic energy stored in the long, thin sample is much higher than the energy dissipated through crack propagation and the creation of new fracture surfaces through the sample cross-section.

By contrast, the initial conditions of stress-intensity measurements by means of CT-specimens are well defined by precracks introduced at the test temperature. The use of a suitable specimen geometry and stress distribution inside the CT-sample ensures that the crack shows a controlled propagation and stops within the

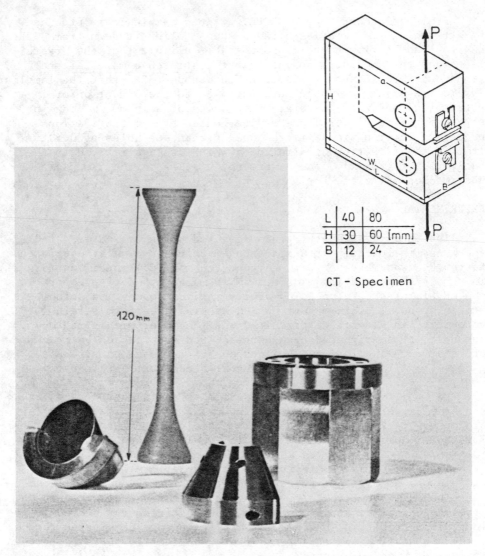

L	40	80
H	30	60 [mm]
B	12	24

CT – Specimen

Fig 2.   CT-specimen and a cylindrical rod specimen with sample holder.

Table 1. Differences in σ_B and K_{1c} measurements.

	σ_B	K_{1c}
sample	cylindrical	CT (compact tension)
load	uniform	inhomogeneous
destruction	total	controlled
precracked	no	yes

sample length according to Griffith's criteria.[7] This crack stop traces a characteristic line in the fracture surface, called <u>arrest line</u>. It is used for measuring the crack length and the fracture area.

The differences of the two measurements are listed in Table 1.

MECHANICAL PROPERTIES AND RESULTS
Epoxide Resins

The properties characterising the materials (Table 2), being v, ρ, σ_B, E, G and K_{1c}, are within 30% for all four epoxide systems under investigation. This shows that these material constants are nearly independent of cross-link distances or of the chemical structure of the epoxide resins. Only for the energy of fracture γ, was a clear dependence on chemical structure found. The stress-strain behaviour is linear which means that epoxide resins follow Hooke's law until fracture.

Polyethylene

The most notable feature is the big difference in the critical stress-intensity factor K_{1c} and the energy of fracture γ for HDPE and EP, whereas the other mechanical constants - except tensile strength σ_B - are approximately equal for HDPE and EP. We found for unstretched polyehtylene that:

K_{1c} is on an average 5 times and

γ is 10 times greater compared with EP.

The stress-strain curves also are different. In HDPE one observes not only elastic but also plastic deformation (Fig 3).[8] To obtain more information about these processes, stretched HDPE's were also examined (Fig 4). The samples were cut from stretched materials perpendicular (HDPE$_p$) and lengthwise (HDPE$_\ell$) to the direction of alignment. The parallel samples are difficult to test and only a few measurements are available.

Table 2. Mechanical properties of four EP-systems and unstretched HDPE.

T=4.2K	EP	HDPE (isotropic)		6,8,9
ρ	1.1 - 1.3	.945	(gcm^{-3})	density
ν	.36 - .37			Poisson ratio
E	7300-8200	9700	(MPa)	Young's modulus
G	2200-2500	3200	(MPa)	shear modulus
σ_B	145-200	180	(MPa)	tensile strength
K_{1c}	1.2 - 1.5	7	$(MPam^{-2})$	fracture toughness
γ	100-210	2500	(Jm^{-2})	energy of fracture
ε_B	1.9 - 2.4	4	(%)	tensile strain

Table 3. Critical stress intensity factor K_{1c}, energy of fracture γ and tensile strength σ_B of polyethylene.

	$K_{1c}(MPam^{-2})$		$\gamma(Jm^{-2})$		$\sigma_B(MPa)$		
Temp.	2K	4.2K	2K	4.2K	2K	4.2K	77K
$HDPE_{iso}$	$7.2\pm.4$	$6.9\pm.5$	1800 ± 500	2500	170 ± 5	170^+-5	150 ± 8
$HDPE_p$	$4.6\pm.5$		1060 ± 300		104 ± 10		100 ± 10
$HDPE_\ell$	>14.				274 ± 20		240 ± 20

In Table 3 the results of measurements on stretched and unstretched HDPE at low temperatures are given.

DISCUSSION

I) Tensile strength, stress intensity factor and energy of fracture.

What is the origin of the discrepancy between two materials having the same tensile strength but almost a factor of ten difference in their energies of fracture and a factor of five difference in their critical stress intensity factors?

Initially, one might expect that a material with a higher K_{1c} would show a higher σ_B. But the results for several EP's and HDPE's indicate otherwise.

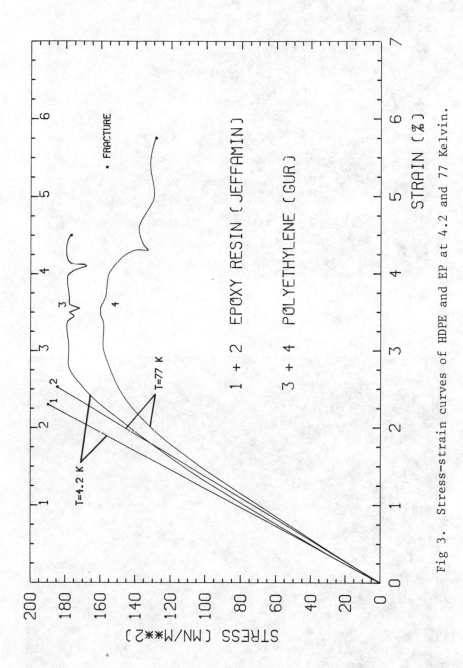

Fig 3.   Stress-strain curves of HDPE and EP at 4.2 and 77 Kelvin.

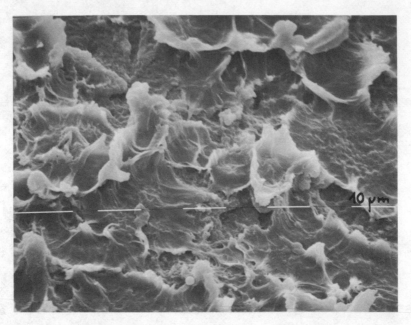

Fig 4: a) Fracture surface of HDPE (isotropic) (SEM).

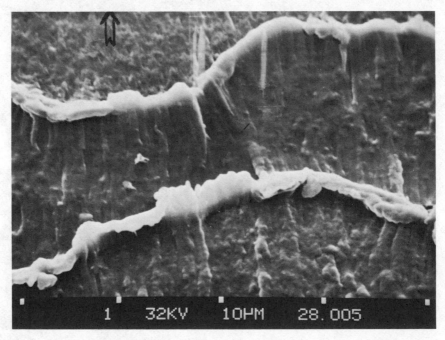

Fig 4: b) Fracture surface of HDPE$_p$.

Ia) One explanation might be the different methods of tensile and crack propagation measurements, precracking being one important difference between the two tests.

Each crack has at its tip a reproducible, and material specific, radius depending on the deformation zone. On tensile loading, a deformation zone is initiated by stress concentration at the crack tip under stable conditions. In this zone the stress near the crack tip is less than the tensile stress. During crack growth heat is generated near the crack tip as a result of the kinetic energy of the broken polymer chains. This forms a moving deformation zone. The size or radius of the deformation zone is a function of the stress concentration, the temperature, the crack velocity, and of the fracture mechanism of the material.

Ib) The deformation zones of the two types of resin differ in size by one or two orders of magnitude.

If a component of stress is high enough there will be - just in front of the crack tip and even at low temperatures - a flow or

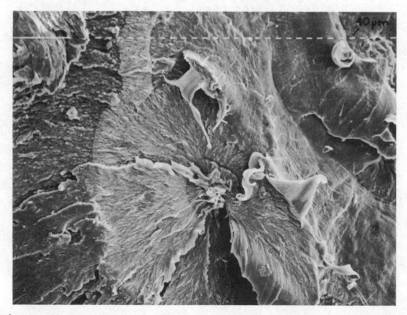

Fig 4: c) Fracture surface of HDPE$_\ell$.
       (At 2 Kelvin, arrow indicates crack direction.)

creep of material forming holes or voids in HDPE, since there is a
large amount of stored energy.

K_{1c} depends on the crack-tip radius and by enlarging this
radius by means of flow or plasticity the point of breakdown will
correspond to a higher load and K_{1c} will become greater.   This is
not possible with the brittle epoxide resin because of the sharp
crack tip and small deformation zone.

In theory one may estimate the deformation zone radius,
according to McClintock and Irwin:[11]

$$r = (1/2\pi)(K_1^2/\sigma_B^2) = r(\text{plane stress})$$

$$r(\text{plane strain}) = r(\text{plane stress})(1-v^2)$$

With K_1 = stress intensity factor (mode 1)

σ_B = tensile strength

K_{1c} = critical stress intensity factor or fracture toughness

$K_{1c} = K_1$ at crack propagation

if $K_1 = K_{1c}$ than becomes $r = r_{max}$ and that is the same as:

$$r_{max} = (1/2\pi)(E\gamma/\sigma_B^2) \text{ (for plane stress)}.$$

In Table 4, r_{cal} is calculated using the above formula and
r_{obs} is the observed radius of the arrest lines (see appendix)
taken from the fracture surface.

Table 4.   Radius of deformation zone in comparison with arrest lines.

	EP		HDPE	
	r_{cal}	r_{obs}	r_{cal}	r_{obs}
4.2K	.009	.005	.25	.1-.3 (mm)
77K	.03	.04	1.0	> 1 (mm)

II)   Temperature dependence.

σ_B shows only a small temperature dependence, whereas r and
K_{1c} vary with temperature.   The size of the arrest lines is a

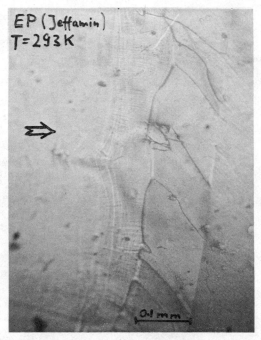

Fig 5: a) Arrest lines of EP (Jeffamin) at 293 Kelvin.

Fig 5: b) Arrest lines of EP (Jeffamin) at 77 Kelvin.

function of the temperature also (Fig 5) and its shape is similar to a cross-section of the deformation zone in the fracture plane.[11] Both values r_{cal} and r_{obs} have the same tendency for temperature dependence and are of the same order of magnitude.

III) Different fracture mechanisms for EP and HDPE.

It is clear that the fracture mechanism in an amorphous material like EP is not as complex as in HDPE which consists of two different phases. The HDPE matrix is also amorphous but there are lamellar crystallites embedded. Even if they are formed from the same chains, they have different mechanical properties.

Examples for possible mechanism before and during fracture:

1) partial alignment of chains and lamellas by stress (HDPE);

2) gliding along crystallite surfaces (HDPE);

3) disentanglement of the amorphous areas between lamellar crystallites (HDPE);

4) defibrillation or breaking of chains (HDPE and EP).

SUMMARY

The only functional relationship between the tensile strength and the critical stress intensity factor is the formula mentioned above. The measurements on CT-samples may be explained by the different deformation zone radii for EP and HDPE. On the other hand this formula gives no explanation for the experiments with rod samples which lead to the same tensile strength of both materials until defects of about 1 mm can be found in the unstretched HDPE.

APPENDIX

Remarks on the nature of the <u>arrest lines</u>.

1) Immediately before a crack stops, the crack velocity slows down. The very sharp crack tips of EP's split into different planes. The reverse is true after crack start: a small distance is needed to recombine the split fracture surface into one plane (Fig 5). In this conception the shiny fracture surface may be associated with fast crack propagation and split surface to slow crack velocity. The domain of slow crack propagation depends on the temperature and has the same magnitude as the calculated deformation zone.

2) There is another phenomenon belonging to EP's as well as to HDPE's: on the SEM pictures of Fig 6 one recognises parts of a

Fig 6: a) Arrest lines of HDPE at 2 Kelvin having roll character (SEM).

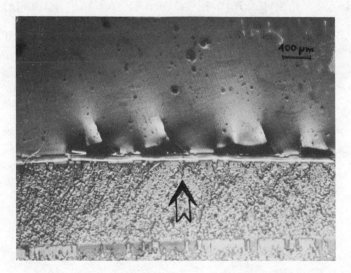

Fig 6: b) Arrest lines of EP at 4.2 Kelvin having roll character (SEM).

roll. This is like a picture of a frozen deformation zone at the crack tip. Deformation zones exist even at crack stop because the load does not drop to zero.[6] The crack splitting away from the normal fracture area calls for mixed stress modes in this zone. This should be a point for further theoretical investigations.

These are the two different types of arrest lines.

REFERENCES

1. CIBA-Geigy systems: X 186/2476/HY 905; CY 221/HY 956;
      CY 221/HY 979; MY 790/Jeffamin D 230.
2. BASF Polyethylene 3428 and 3429 extruded material.
3. Hoechst (Ruhrchemie): GUR pressure sintered material.
4. Kanig G, Kolloid-Z.u.Z. Polymere 251, pp. 782-783 (1973).
      Kanig G, Progr. Colloid & Polymer Sci. 57, pp. 176-191 (1975)
5. Hartwig G, Wuechner F, Rev. Scient. Instr. 46, Nr. 4 (1975),
      p. 481.
6. Kneifel B, Nonmetallic Materials and Composites at Low
      Temperatures (1979) pp. 123-129; Plenum Press, New York.
7. Griffith A A, Phil. Trans. Roy. Soc. London, A, vol. 221,
      pp. 163-198 (1921).
      Griffith A A, Proc. 1st Int. Congr. Appl. Mech., Delft (1924),
      p. 55.
8. Hartwig G, Nonmetallic Materials and Composites at Low
      Temperatures (1979) pp. 33-50; Plenum Press, New York.
9. Hartwig G, Progr. Colloid & Polymer Sci. 64, pp. 56-67 (1978).
10. Hartwig G, Wuechner F, Materialpruefung 18 (1976), Nr. 2 Feb.,
      pp. 40-44.
11. McClintock F A, Irwin G R, ASTM STP Nr. 381, Philadelphia
      (1965).
12. BBC (Brown, Boveri & Cie) Druckschrift: D NG80761 D and D
      NG60872 D.

FATIGUE TEST OF EPOXIDE RESIN AT LOW TEMPERATURES *

S Nishijima, S Ueta and T Okada

Department of Nuclear Engineering
Osaka University, Suita
Osaka 565 Japan

1. INTRODUCTION

In the various large scale projects concerned with energy
resources, superconducting magnet technology becomes increasingly
important; for example, in magnetic confined fusion reactor SC
magnets are believed to be necessary from power balance. In the
case of the Tokamak reactor both toroidal and poloidal magnets
should be constructed as SC magnets. The former type magnet is to
produce a large and intense magnetic field (DC mode), but the latter
is to generate a pulsating field (AC mode). In these magnets, the
materials are exposed to complex stress/strain states in a complicated dynamic and time dependent manner.

The principal components of SC magnets are (i) superconductor,
(ii) insulator, and (iii) structural material. Among these,
superconductor has received much attention. The insulating or
potting material, however, has received comparatively little
attention. The roles of insulating materials in magnets are as
follows: (i) insulation of turn to turn or layer to layer etc,
(ii) restriction of the wire movement, (iii) spacer which provides
cooling channel for liquid helium, (iv) thermal insulation, and
(v) structural material as bobbin or bolt etc. These materials
must have appropriate electrical and mechanical properties at
cryogenic temperatures. These two are not independent of each other,
eg it has been reported that mechanical crack gives serious effects
to insulating properties.[1] Careful examination of fracture
mechanism is, therefore, needed.

* This work is supported in part by the Grant in Aid for Scientific
Research No 504533, Ministry of Education in Japan.

139

Among many kinds of mechanical properties, fatigue resistance may be one of the critical design parameters in superconducting magnets operated in periodic driving mode.[2,3] The investigation of fatigue characteristics involves the degradation or the destruction of the material under varying stress. In general, material is destroyed at much lower stress levels than the static breaking stress. In many structures or machines it is believed that total failure is often triggered by material fatigue.[4] The fatigue behaviour is particularly important in pulsed magnets. For example, according to the design of JT-60 of Japan Atomic Energy Research Institute,[5] its poloidal magnet is magnetically excited every 400 seconds. This means the magnet will receive periodic stresses for approximately 10^6 cycles, if 20 years is assumed as the life of this facility.

The purpose of the present study is to obtain experimental data at LHeT, to examine the temperature dependence of fatigue properties and to provide reference data for magnet design.

2. EXPERIMENTALS AND RESULTS

2.1 Sample

The test material is a commercially available epoxide resin (6861-1)* which has been used in the SCM as an insulator. It includes as chief ingredient (6861-2) the Bisphenol-A resin and a curing agent (8879-2) of polyethanol complexes of titanium with a small amount of diluent (phenyl glycidyl ether).

Curing conditions are 150°C for 12 hours followed by slow cooling (0.5° C/min) to room temperature (RT). The samples are cut into cylindrical shapes 4.0 mm diameter and 30 mm in length, and are conditioned at approximately 5°C in a desiccator for more than one month.

2.2 Flexural Test

A flexural test is made in order to get basic information applicable to fatigue testing. It is established that organic materials show deformation rate dependence of mechanical properties.[6] The change in mechanical properties with deformation rate should be clarified before carrying out a fatigue test, because the fatigue test is usually made at a high frequency in order to shorten the period required for testing.

The geometry of the sample in this experiment is shown in Fig 1. The method used is fundamentally a three point bending test.

---

* This is kindly supplied by Ryoden Kasei Co Ltd.

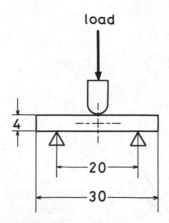

Figure 1. Diagramatic illustration of test specimen. The method is fundamentally a three point bend test

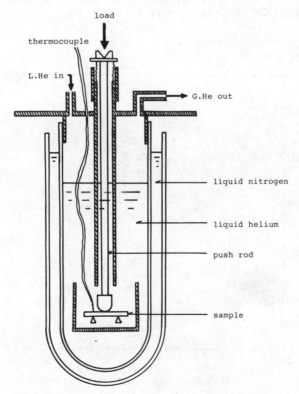

Figure 2. Schematic illustration of the experimental apparatus in liquid helium

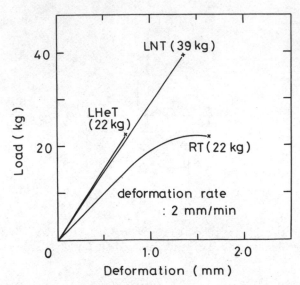

Figure 3. Typical load-displacement curves at RT, LNT and
LHeT obtained with deformation rate of 2 mm/min

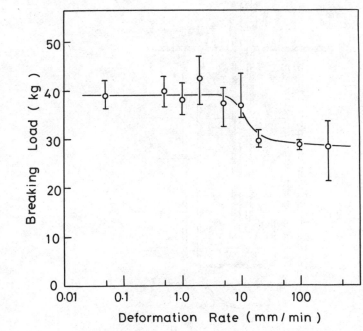

Figure 4. Deformation rate dependence of breaking load at
LNT

A punch applies force to the specimen, which is supported by two
knife edges with 20 mm span. A Tensilon Universal Testing Instrument
Model (UTM-1) is used in this experiment. The flexural test is
performed at RT: 13∿20°C, LNT (liquid nitrogen temperature), and
LHeT (liquid helium temperature). The experimental apparatus for
the flexural test at LHeT is represented schematically in Fig 2.

Typical load-displacement curves obtained with a deformation
rate of 2 mm/min at RT, LNT, and LHeT are shown in Fig 3. At RT,
the specimen is broken after yielding; on the contrary at both LNT
and LHeT, yielding is not observed. The elastic constant increases
with decrease in temperature. For the breaking stress, however, a
temperature decrease does not always increase the breaking stress,
the largest breaking stress being measured at LNT.

Fig 4 shows the deformation rate dependence of breaking load
at LNT. The breaking load shows an abrupt decrease approximately
at 10 mm/min deformation rate and decreases gradually with increased
deformation rate. This phenomena is previously reported by
Nishijima and Okada[7] and can be recognised in terms of a relaxation
process. The breaking load at 2 mm/min deformation rate is defined
as the "static breaking load" because the breaking load appears to
be constant below 2 mm/min deformation rate. The static breaking
load is also defined as the breaking load at 2 mm/min at RT and
LHeT for convenience.

2.3 Fatigue Test

Concerning fatigue, the design parameters for cryogenic
temperatures are apt to be based on the fatigue data at RT since a
temperature decrease gives a higher elastic modulus and breaking
stress to materials.[8] There are some reports that the breaking
stress at LHeT is smaller than that at LNT in both tensile
flexural tests.[9] Because of this the exact fatigue data at cryogenic temperatures is required.

A self-made fatigue testing apparatus is used, shown schematically in Fig 5. The geometry of the specimen in the fatigue
test is the same as that for the flexural test (Fig 1). A cam
moves a pivoted bar up and down and a weight is suspended at the
end of the bar. The other end of the bar pushes a steel ball placed
on a push rod which transmits a pulsating load to the specimen.
The frequency of the cam rotation is 2 Hz and hence that of the
pulsating load is 2 Hz. The size of the pulsating load is adjusted
by changing the weight.

The shape of the load cycle is represented in Fig 6 and is
not exactly sinusoidal. The maximum value of the load is termed
the "fatigue load" in this case, although the average load is
often referred to as the fatigue load.

The results of the fatigue test are shown in Fig 7, which demonstrates the load-endurance diagrams at RT, LNT, and LHeT. These diagrams are divided into two different regions. In region I, the fatigue life increases exponentially with a decrease in fatigue load where the diagrams have a straight line with negative slope. In region II, infinite load cycles are necessary for the sample to fracture where the diagram is parallel to the fatigue life axis and the load level is called "fatigue limit". The design load should naturally be under the fatigue limit. The fatigue limit at LNT is approximately twice that at RT or LHeT and this is expected from the results of the static flexural test. 10^6 cycles is chosen as the ultimate fatigue cycle because the superconducting magnet used in the fusion reactor will be subjected to a pulsative stress of the order of 10^6 cycles.

3. DISCUSSION

The probability of crack initiation during the fatigue test may be defined as the slope of the load-endurance diagram (Fig 7). We are unable to compare the probability at different temperatures directly because the load for crack initiation is thought to vary with temperature and this may be observed as follows. From the abrupt fall of stress after fracture in a load-displacement diagram, the propagation of a crack is rapid, and hence the external applied force would not be necessary to propagate the crack after initiation. Because of this the breaking load is presumed to be the load for crack initiation and the breaking load changes with temperature as reflected in Fig 3. In order to investigate the change in the probability of crack initiation with fatigue load, the normalised fatigue load in terms of static breaking load is employed as the ordinate in Fig 8. Some experimental equations have been proposed to approximate the stress-endurance diagrams.[10] Among them the following equation seems to show a good approximation of the experimental results.

$$S = A - M\log N$$

where S is normalised fatigue load, N is fatigue life, and both A and M represent certain constants. This equation may be applied to a higher load level than the fatigue limit.

The constant M represents the probability of crack initiation against fatigue load, in other words, the sensitivity to repeated loading. The value of parameter M increases with temperature as shown in Table 1 and thus it can be concluded that temperature decrease makes the sample less sensitive to cyclic loading, normalised to the static breaking load.

The parameter A should be less than one and it represents the reduction of breaking load caused by rapid fatigue deformation rate.

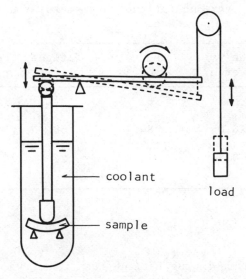

Figure 5.   Principles of operation of fatigue testing apparatus

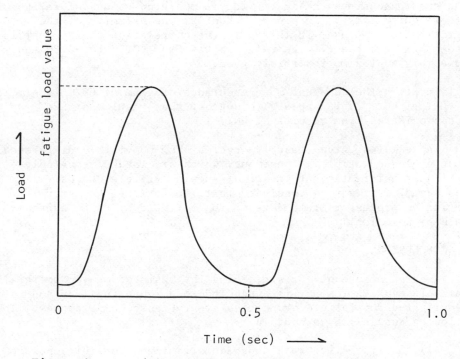

Figure 6.   Load/time characteristics in fatigue test

Table 1. Various fatigue parameters at different
temperatures

temperature	the value of M	the value of A	fatigue limit	correlation coefficient
LHeT	7.27	83.7%	54%	−0.97
LNT	8.05	84.1	58	−0.87
RT	9.36	73.9	27	−0.80

The deformation rate in the fatigue test is much higher (several
hundreds) and hence the breaking load should decrease compared with
the static breaking load. The value of A at different temperatures
is also presented in Table 1.

There has been little confirmation of functional formulae for
load-endurance since the scattering in measured values of fatigue
life is quite large. In order to confirm the validity of the
approximation employed in this work the correlation coefficient is
also shown in Table 1. The value ranges from −0.97 to −0.80 and
shows this approximation fairly good.

In Fig 8 the shape of load-endurance diagram at LHeT is similar
to that at LNT. This appears to suggest that results at LHeT may
be approximated from results at LNT.

The samples which stand 10^6 cycles show no visible difference
to an un-loaded sample. Specimens, which are loaded repeatedly
until just before the fatigue life, show no visible cracks.
Consequently, it may be concluded that, in fatigue, the crack
initiation process is rate-determining followed by rapid adiabatic
crack propagation.

4. CONCLUSIONS

A cured Bisphenol-A resin, being the basis of many commercially
available epoxide resins, is subjected to a repeating load and is
investigated for fatigue properties at cryogenic temperatures. The
following conclusions are drawn.

(1) A deformation rate dependence is found for this epoxide
resin in the flexural test and hence the breaking load at the first

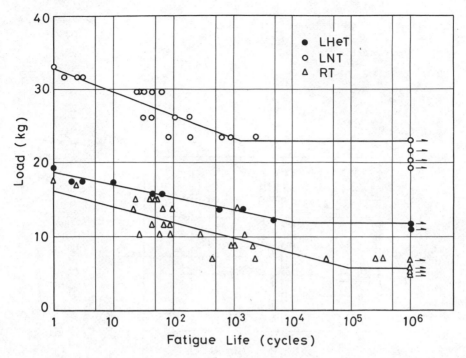

Figure 7. Load-endurance diagram at RT, LNT and LHeT

loading in fatigue test is lower than the static breaking load at
RT, LNT and LHeT.

(2) Temperature decrease reduces the specimen sensitivity to
fatigue loading, even though the absolute breaking load level on
the load-endurance diagram at LNT is always larger than at LHeT.

(3) The load-endurance diagram obtained at LNT is similar to
that at LHeT if the load axis is normalised in terms of the static
breaking load.

(4) The crack initiation process is the rate-determining
process in fatigue at cryogenic temperatures.

5.   ACKNOWLEDGEMENTS

The authors are grateful to Dr J Yamamoto, Mr T Tsuji,
Mr Y Wakisaka and Mr H Makiyama of Low Temperature Centre of Osaka

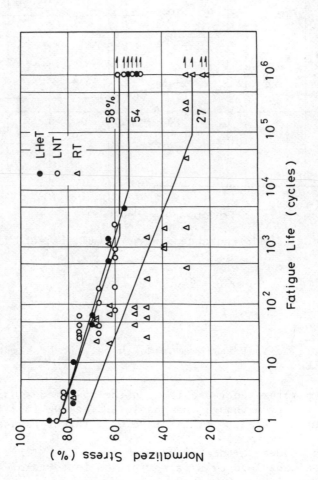

Figure 8. Normalized load-endurance diagram at RT, LNT and LHeT : (Normalized stress (%)) = fatigue load)/(static breaking load) x 100

University for their help in the experiment using liquid helium.
They would like to thank Mr T Nakamura of Ryoden Kasei Co Ltd for
supplying the epoxide resin.   They are also grateful to
Dr T Hagihara of Osaka Kyoiku University for various stimulating
suggestions.

REFERENCES

1.   D C Phillips, AERE-R-8923 (1978).
2.   R L Tobler, D T Read, J. Comp. Mat. 10 (1976) 32.
3.   E L Stone, L O El-Marazki, W C Young, Nonmetallic Material
          and Composites at Low Temperature, ed. A F Clark et al,
          Plenum Press N.Y. (1979) 283.
4.   Y Kawata, Y Matuura, et al, "Material Test" Kyouritsu Pub. Co.
          Tokyo (1975).
5.   Fusion Reactor System Laboratory, JAERI-M 7200 (1977).
6.   M Suzuki, S Iwamoto, Documents of the Committee of FRP Division,
          Society of Material Science of Japan (1978).
7.   S Nishijima, T Okada, Cryogenics 20 No 2 (1980) 86.
8.   K Nakagawa, H Kobayashi, "Fundamentals of Fatigue Metal"
          Youkendo (1975).
9.   B Fallou, MT-5 Roma (1975) 644.
10.   T Yokobori, "Strength, Fracture and Fatigue of Materials"
          Gihoudo Pub. Co. Tokyo (1976).

LAP TEST OF EPOXIDE RESIN AT

CRYOGENIC TEMPERATURES

S Nishijima and T Okada

Department of Nuclear Engineering
Osaka University, Suita
Osaka 565 Japan

INTRODUCTION

The reasons why the expected performance of superconducting magnets is not easily achieved are because of instabilities such as training and degradation. Impregnation or potting techniques have been employed in order to modify such instabilities. Epoxide resin has been selected in this work from the range of impregnating materials (eg woods metal,[1] epoxy[2] [5] and wax[6]) because epoxide resins have been shown to possess excellent adhesive strength and satisfactory mechanical properties at cryogenic temperatures.

Epoxide resins, however, also have disadvantages such as large thermal contraction in comparison with that of metals. Because of the difference in thermal contraction the epoxide resins may not perform satisfactorily when they are used with metals or as adhesives to fix metals at cryogenic temperatures. The reasons for magnet quench are reported to be due to wire movement and/or frictional heating for poorly bonded conductors,[7] [8] or mechanical cracking[9] at the boundary between superconductors and potting material, originating from thermal stress or Lorentz forces. From the viewpoint of importance of adhesive strength, the mechanical behaviour of epoxide resins is investigated in this work in order to give a basis for potting material selection.

This study is divided into three parts. The first is the race track coil test, in which coils are stretched at low temperatures in order to observe the boundary behaviour under stressed conditions. The second is a compression test of epoxide resins to determine the mechanical properties of the basic material. The third is a lap test to obtain basic information about the mechanical behaviour of the potting materials as the adhesive which fixes the wires to each other.

As mentioned above, bonding defects between conductor and potting material may trigger wire movement and heat generation and thus we should take measures to avoid these perturbations. There should be two ways of achieving this, one is to prevent the initiation of defects.[10] [18] The other is to control heat generation even under the existence of defects. Concerning the former there are few studies [19] [22] about the boundary behaviour. On the latter, the study of coefficient of friction between metals and insulating materials has been started. [8] [9] These two studies supplement each other.

The purpose of this investigation is to get basic information about the initiation of defects, which may control magnet degradation, and to clarify the problems concerning the stability of superconducting magnets.

EXPERIMENTAL AND RESULTS

Stretch Test of Race Track Coil

Potted race track coils made of Cu, NbTi or NbTiZrTa wire are stretched at RT (room temperature), LNT (liquid nitrogen temperature) and LHeT (liquid helium temperature). The copper coil is useful for checking the boundary behaviour in large deformation because copper shows a larger breaking strain than the superconductors.

Figure 1 gives the specifications of the coils. The potting resin is epoxy (6861-1) of which the chief ingredient is bisphenol-A supplied by Ryoden Kasei Co Ltd. The coils are potted using this resin for 6 hours at 55°C in vacuum, followed by heating to 150°C for 12 hours to cure the resin.

Coil No 1 is stretched at RT as a preliminary step. GFRP rods which have an outer diameter the same as the inner diameter of the race track coils at the ends are used to apply load to the coils. An Instron Universal Testing Instrument Model 1125 is used in this experiment. The GFRP rods contain SUS-axle in order to reduce the deflection of itself.

Coil No 2 is made from copper stabilised NbTi. This coil was cooled to LNT rapidly and many cracks transversing the coil were present; after, the coils were cooled gradually to LNT with approximately 3 hours which is enough to prevent the initiation of observable cracks. Number 2 coil was stretched at LNT at the tensile speed of 0.5 mm/min. Figure 2 shows the force-displacement curve of the coil. In this figure the abscissa represents crosshead

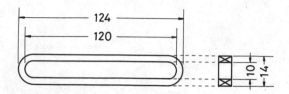

Coil Number	Conductor	Diameter	Cross Section	Turn	Tension
1	Cu	0.5 µm	2 x 2	8	1.6 kg/wire
2	NbTi	0.28	2 x 4	49	1.6
3	Cu	0.5	2 x 4	17	1.6
4	NbTi	0.28	2 x 4	46	0.8
5	NbTiZrTa	0.35	2 x 4	40	1.6

Figure 1.   Shape and specifications of tested coils.

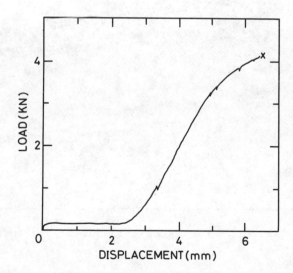

Figure 2.   Force-displacement curve of coil No.2.

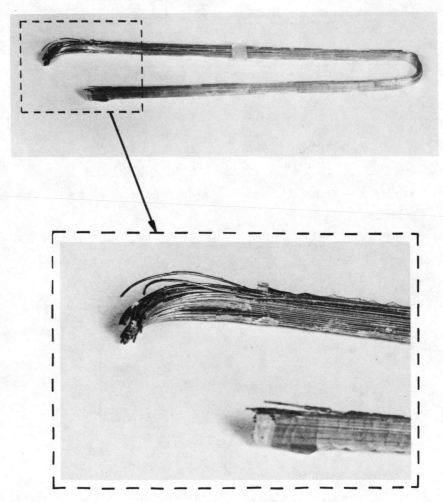

Figure 3.   Appearance of tested coil No. 2: Fractured part of
the coil.

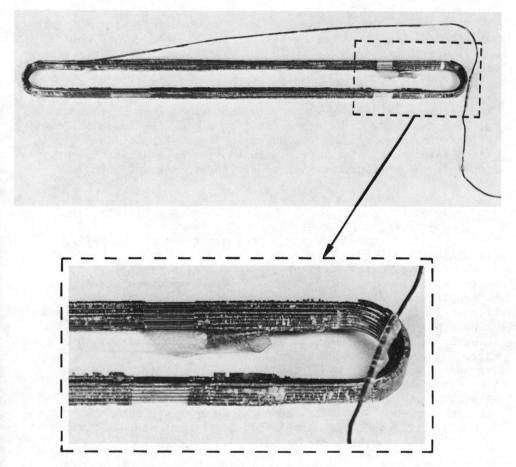

Figure 4. Appearance of tested coil No.3 : Formal peeled off
parts of the coil with potting epoxy.

displacement. Stress changes are found during loading and these are attributed to cracking of potting material and/or rearrangement of wires. This tested coil is presented in Figure 3. Potting material is peeled off in some places, and is almost missing at the curved parts where stress concentration is greatest. The difference of elastic modulus between resin and wire appears to cause shear stress at the boundary leading to 'peeling-off' of potting material. At the straight part many new cracks are initiated across the coil.

In Figure 4 the photograph of·coil No 3 is represented. This was deformed about 23 mm of crosshead displacement. This coil appears seriously stretched at the straight part. The shear stress between wire and potting material causes peeling of insulator from the copper, and the wires show bare copper surface in some places.

The pre-tension of coil No 4 is approximately half that of coil No 2 and hence coil No 4 shows larger breaking deformation. The load-displacement curve is shown in Figure 5. The lack of potting material at the curved parts and the initiation of mechanical defects in the potting material is similar to that of coil No 2.

Coil No 5 is constructed using NbTiZrTa. This coil carries the superconducting current at the LHeT under 4 T of external field. The field is perpendicular to the current of the straight part. Because of this, Lorentz forces act to cause the coil to rotate; in other words, this coil is under more severe stress than the usual experimental configuration. A coil supporter is arranged to prevent coil rotation. But the coil is compressed partly between the supporter resulting in distortion (Figure 6). The influence of external forces on training of this type of coil will be reported in the separate paper.[23] It is clear that mechanical defects are caused in the boundary between wires and potting material when a potted coil is deformed or is subjected to forces because such deformations or forces would cause non-uniform deformation and /or stress in a coil. These defects might provide training and/or degradation to the coil. Consequently, a compressive test of the resin and lap shear test are needed to investigate both the process of defect initiation and the relation between mechanical properties and adhesive strength of the resin.

Compressive Test

A compressive test of resins which are used in lap test is carried out at RT, LNT and LHeT in order to obtain the mechanical properties of these resins. Small cylindrical test pieces of 1.5 mm height and 1 mm diameter are prepared. The compressive speed is 0.5 mm/min. Two sorts of epoxies are supplied for this work. The one (hereinafter called "sample A") is the resin used in race track coil test. Both resins have the same chief ingredient of bisphenol-A but have a different hardener. The hardener of

sample A is polyethanolamine complexes of titanium and the other
(named "sample B") contains aliphatic amine as a hardener. Sample A
and Sample B are cured at 150°C for 12 hours and at 120°C for 3 hours,
respectively. These conditions are thought to give the best cure.
Figure 7 represents typical stress-strain curves at RT, LNT and
LHeT. Compressive deformation curve of sample A and B both show
yielding at RT and LNT. At LHeT they both fracture without yielding.
Results of compressive test such as breaking stress, breaking strain,
elastic modulus and yield stress are summarised in Figure 8. The
experimental results, shown in Figure 8, are the average of five
measurements. It appears that there are few differences between
mechanical properties of sample A and those of sample B.

Lap Shear Test

    Lap shear test is carried out since it is considered that
adhesive strength between wires and potting material should also be
examined for impregnating materials. The shape of the specimen is
shown in Figure 9. Two copper strips of width 5 mm, length 25 mm
and thickness 0.1 mm are placed together with a 5 mm overlap. An
epoxide resin is spread over the lapped part of the copper strips.
These strips are pressed at about 20 g/cm^2 and cured. The samples

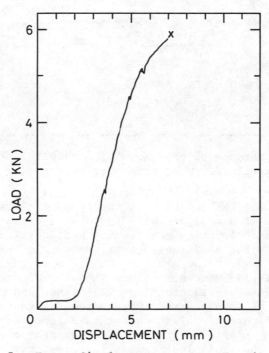

Figure 5. Force-displacement curve of coil No.4.

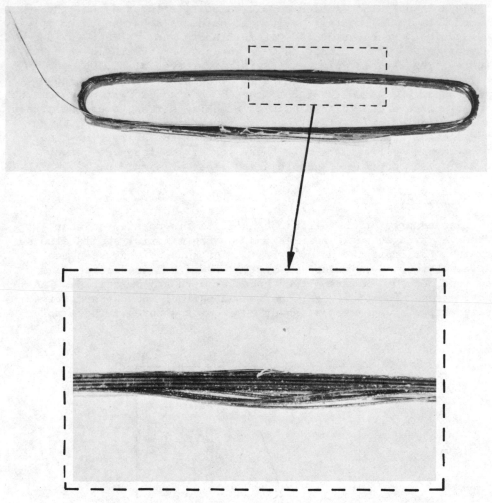

Figure 6. Appearance of coil No. 5 : Separation of winding due
to electro-magnetic force by transport current.

are extended at RT, LNT and LHeT at the speed of 2 mm/min. Two
kinds of epoxide (A and B) are used in this experiment. The thickness of the epoxide layer is measured by sectioning and is about
0.03 mm. The roughness of these copper strips is 2-4 μ.

    The load-displacement curves are shown in Figure 10. The
abscissa represents displacement in millimetres and degree of strain.
The ordinate shows load and the average shear stress per unit lap
area. In this figure the displacement is defined as the remainder
after subtracting displacement of the copper from the total specimen
displacement. Hence the displacement in this figure may be
considered that of lapped part. The temperature decrease brings

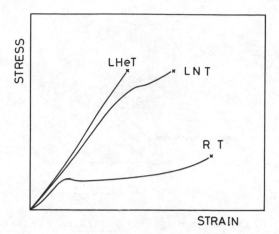

Figure 7. Typical stress-strain curves at RT, LNT and LHeT
in compressive test.

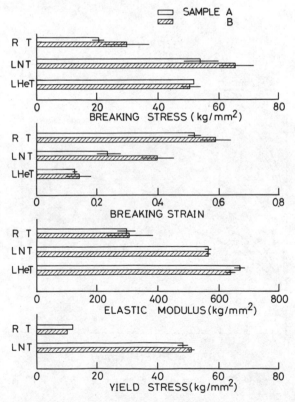

Figure 8.   Comparison of various mechanicl properties of sample
A and B at RT, LNT and LHet. The line segments on the top of
bar graph mean error bars.

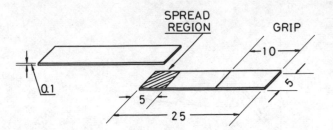

Figure 9. Shape of the specimen used in lap shear test.

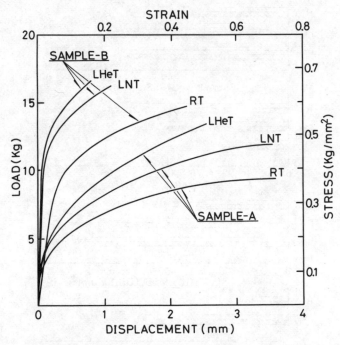

Figure 10. Load-displacement curves at RT, LNT and LHeT in lap shear test.

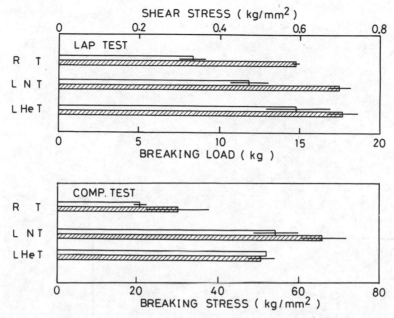

Figure 11. Comparison of breaking stress of lap and compressive test. Unhatched and hatched bars show sample A and B, respectively.

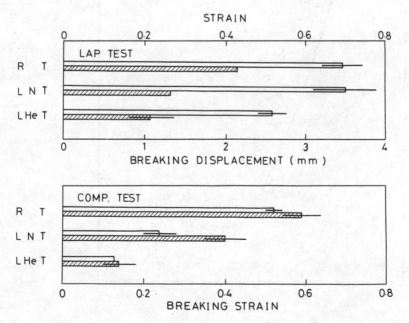

Figure 12. Comparison of breaking strain of lap and compressive test. Unhatched and hatched bars show sample A and B, respectively.

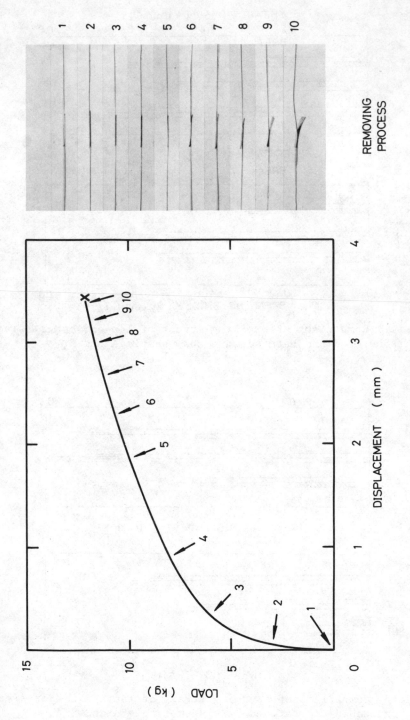

Figure 13. Peeling process of lapped specimen.

lower breaking displacement and higher breaking load in both cases
of A and B. The comparison reveals that the sample B shows higher
breaking stress and lower breaking strain than sample A.

The results of lap test and compressive test are compared in
Figures 11 and 12, which represent breaking stress and breaking
strain, respectively. It is found that the results of lap test do
not agree with that of compressive test in some cases.

Concerning sample A, to investigate the peeling processes of
the specimens the deformation is stopped at a certain load level
during the test at LNT. The relationship between deformation mode
of the sample (shown by photographs) and load-displacement curve is
shown in Figure 13. The peeling starts from both ends of lapped
area and meets at the centre. This means that the stress in the
adhesive layer is different from point to point and distributes in
the form of smallest at the centre and largest at the ends. Because
of this yielding does appear in lap test at LHeT, though it is not
seen in compressive test at LHeT.

Figure 14 presents sectional views of specimens with the
traced figures. The shadowed portions mean adhered places of the
epoxy and the others indicate the parts peeled off, that is the
epoxy adheres to the counterpart of the strip. The breaking modes
of such composites are generally divided into the following four
types;

(1) cohesive failure (failure of the adhesive)

(2) interface failure (adhesive failure)

(3) coexistence cohesive and interface failure

(4) adherend failure (failure of the adherend material).

Figure 14 represents that coexistence of cohesive and interface
failure at RT, while interface failure takes place at LNT and LHeT
in the case of sample A. In the case of sample B coexistence of
cohesive and interface failure occurs at each temperature. A
large cohesive force makes the resin hard and brittle resulting in
low ultimate strength of the composite because the large cohesive
force does not allow the resin to relax its stress and causes
large stress concentrations. On the contrary small cohesive force
results in cohesive failure though the composite can resist impact
stress because of stress relaxation. Consequently, the highest
adhesive force is demonstrated when the force for the interface
failure is similar to that for cohesive failure. Sample B presents
higher adhesive strength since sample B shows the fracture mode of
coexistence cohesive and interface failure, even at cryogenic
temperatures, in other words, under residual stress condition
caused by differential thermal contraction at the boundary. At

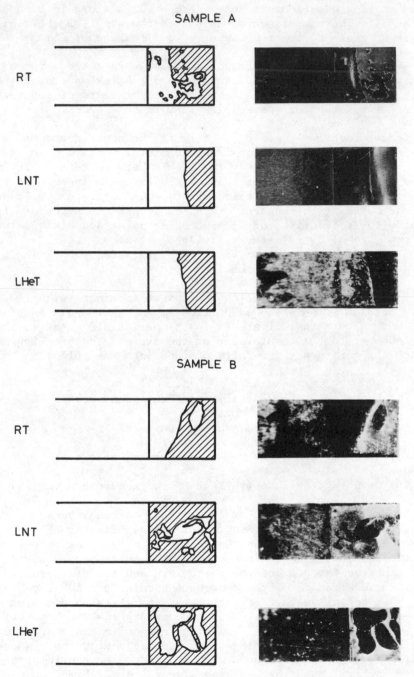

Figure 14. Views of fractured surface in lap test.

this point sample B is better than sample A, even though these two samples show similar mechanical behaviour.

CONCLUSIONS

Stretch tests of potted coil and lap shear tests have been made to reveal the following results.

(1)   The boundary behaviour between wire and potting material may control the performance of magnets in practice.

(2)   Materials used in composite form such as impregnating materials should be selected after careful consideration not only of the mechanical properties themselves but also of adhesive strength.

(3)   The cohesive force does not always reflect the adhesive strength.

ACKNOWLEDGEMENTS

The authors are grateful to Dr J Yamamoto, Mr T Tsuji, Mr Y Wakisaka and Mr H Makiyama of Low Temperature Centre of Osaka University for their help in the experiment using liquid helium. They would like to thank Mr T Nakamura of Ryoden Kasei Corp. for supplying epoxide resins. They are also grateful to Dr T Horiuch of Kobe Steel Ltd for supplying the superconductors.

REFERENCES

1.   K Kuroda, Cryogenics 15 (1975) 675.
2.   V W Edwards, C A Scott, M N Wilson, IEEE Trans. Mag-11 (1975) 532.
3.   O P Anashkin, V A Varlakhin, V E Keilin, A V Krivikh, V V Lyikov, IEEE Trans. Mag-13 (1977) 673.
4.   S G Ladkany and Z L Stone, Nonmetallic Material and Composites at Low Temperatures, ed. A F Clark et al, Plenum Press NY (1979) p377.
5.   Z N Sanjana and M A Janocko, ibid (1979) p387.
6.   P F Smith, B Colyer, Cryogenics 15 (1975) 201.
7.   M Wilson, IEEE Trans. Mag-13 (1977) 440.
8.   Y Iwasa, R Kensley, J E C Williams, IEEE Trans. Mag-13 (1979) 20.
9.   Y Iwasa, R Kenslay, L E C Williams, IEEE Trans. Mag-13 (1979) 36.
10.   G Hartwig, IEEE Trans. Mag-11 (1975) 536.
11.   B Fallou, MT-5 Rome July (1975) 644.
12.   G Hartwig, CEC Boulder Aug. No D-2 (1977).
13.   R P Reed, R E Schramm, A F Clark, Cryogenics 13 (1973) 67.
14.   Van de Voorde, IEEE Trans. Nucl. Sci. 3 (1973) 693.

15. J Thoris, J C Bobo, CEC-6 Grenoble May (1976) 271.
16. E L Stone, L O El-Marazki, W C Young, ICEC Munich July (1978)
     283.
17. M B Kasen, Cryogenics 15 (1975) 327.
18. S Nishijima, T Okada, Cryogenics 20 (1980) 86.
19. G R Imel, P V Kelsey, E H Ottewitte, J. Nuclear Material 85
     and 86 (1976) 367.
20. R H Kernohan, C J Long, R R Coltman, J. Nuclear Material 85
     and 86 (1976) 379.
21. S O Hong, P F Michaelson, I N Sviatoslavsky, W C Young,
     ICMC Boulder, August CA-8 (1977).
22. K J Froelich, C M Fitzpatrick, ORNL/TM-5658 (1976).
23. T Okada, S Nishijima, T Horiuchi, to be submitted to Applied
     Superconductivity Conference, Santa Fe, Sept. 29 - Oct. 2
     (1980).

COHESIVE STRENGTH OF AMORPHOUS POLYMERS AT LOW AND HIGH

TEMPERATURES

M Fischer

Ciba-Geigy AG,
Basel, Switzerland

INTRODUCTION

A force of 6 nN is required to break a carbon-to-carbon bond[1].
It therefore follows that polycarbonate in which an area of 1 mm^2
is traversed by about 10^{12} molecular chains ought to survive a
stress of 2000 MNm^{-2} assuming one chain out of three parallels the
direction of the applied load. However, the tensile yield strength
of polycarbonate[2] is 62 MNm^{-2}. Obviously, the initial assumption
is a fallacy. It can not be the right way to get an answer to the
question – How strong are polymers really?

The flexural test gives an indication of the strength of the
material tested. Figure 1 shows force-deflection curves for a
crosslinked polymer tested at various temperatures. At 100°C, the
polymer yielded under a load of about 150 N. At lower temperatures
the same polymer exhibited greater strength with, for example,
yield occuring at about 230 N at 55°C. At even lower temperatures
there was no yeild at all. At -20°C the sample broke quite un-
expectedly after a promising start.
What would have happened if the samples had not failed prematurely?
Fracture mechanics provides an explanation for failure – flaws
expand under stress. Flaws are always present in polymers and can
not be eliminated. However, this explanation is unable to provide
a reasonable indication of the likely true strength of amorphous
polymers. A simple experiment can be used to eliminate the confusing
effects of flaws. The polymer under investigation is used to bond
aluminium cyclinders to aluminium plates as shown in Figure 2, and
these composites are then subjected to a torsion test. Cracking
is hampered by the unfavourable mode of crack loading (tearing
mode), and crack propagation is blocked by the two bonded aluminium

parts. Consequently, the bonded joint does not fail until the
stresses imposed reach the true strength of the polymer. Of course,
testing composites introduces a new problem: adhesion. Fortunately,
adhesion is adequate in most cases.

In the following, the test method used will be described
briefly, and the results obtained for shear strength will be compared
with the outocme of a simple model based on activated flow. Cohesive strength will be discussed in relation to the molecular
structures of the polymers. Long-term resistance and flexibilization
will be shown to fit the outlined pattern of cohesive strength. It
will then be evident on what factors the strength of glassy polymers
are based.

MATERIALS

The materials investigated were non crystalline organic polymers
at temperatures below their glass transitions, i.e. glassy polymers.
Most of the work was done with epoxide resins. These materials
offer a range of different, crosslinked polymers which can be
produced simply by mixing resins and hardeners of known structure.
Thermoplastic resins were also tested.

EXPERIMENTS

All aluminium cylinders and aluminium plates were etched in
chromic–sulphuric–acid before use. The plates were covered with an
adhesive strip into which 12 mm holes had been punched (Figure 3).
The adhesive mix to be tested was applied in the open areas (Figure 4)
and levelled using a razor blade (Figure 5). The surplus mix was
then removed together with the perforated adhesive tape. Finally,
the cylinders were placed on the 12 mm patches of adhesive (Figure 6)
and the samples were cured.

The test was carried out by placing the plate in a channel cut
in a lever. A torsion couple was connected to each cylinder in turn
by means of a ring spanner as shown in Figure 7. The lever was in
contact with a load cell. The results of five tests were averaged
to obtain a mean. Figure 8 shows shear strength versus temperature
for an epoxide resin cured with an acid anhydride.
It is clear that the shear strength increases as the temperature is
lowered, exactly as in the case of flexural strength. However,
this increase is observed to occur over a much broader temperature
range for shear strength than for flexural strength, i.e. from above
Tg straight down to 77K!

In all, about seventy polymer systems were tested. Not all of
these gave lines as straight as that of the anhydride system in
Figure 9. Systems with pronounced β-relaxation were particularly
prone to show deviations from straight line behaviour at about 220 K.

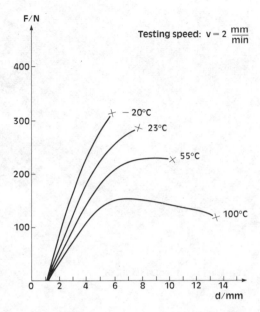

Figure 1.   Force (F) – Deflection (d) – Diagrams of Araldite[®] AV8
            The Yield strength increases as temperature is lowered.

Figure 2.   The polymer is used to bond aluminium cylinders to an
            aluminium plate.

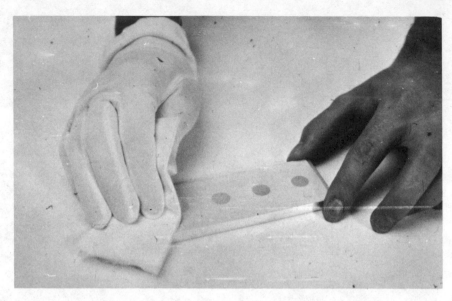

Figure 3.   Pretreated aluminium plates are covered with an adhesive
             strip.

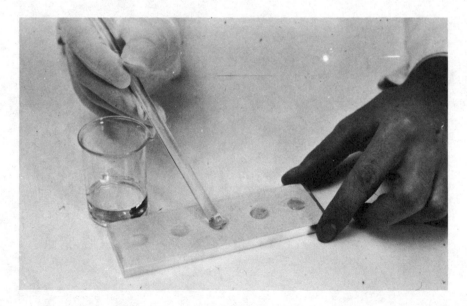

Figure 4.   The adhesive mix is applied over the holes.

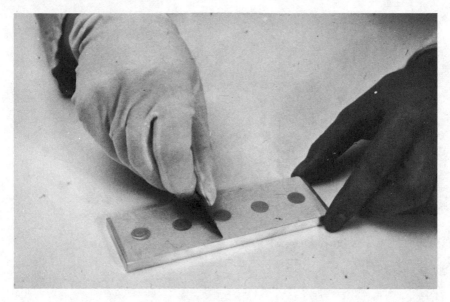

Figure 5.   Surplus mix is removed.

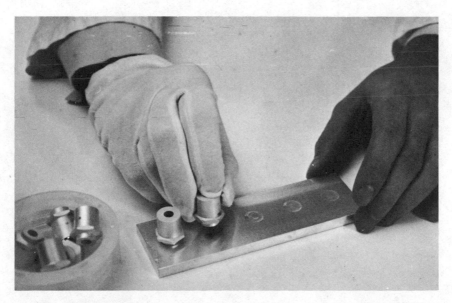

Figure 6.   Pretreated aluminium cylinders are placed on the patches
of adhesive mix.

Figure 7. Testing rig (without temperature cabinet) also used
with liquid nitrogen.

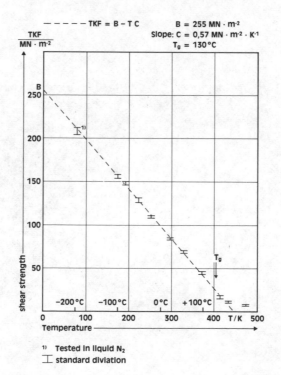

Figure 8. Shear strength versus temperature; AL bonded to AL with
DGEBA + HHTA

Nevertheless, most of the results were found to be in reasonable accordance with the simple relation shown in Figure 8.

$$TKF = B - C \cdot T \tag{1}$$

TKF is shear strength. In case of cohesive failure, this will be called cohesive strength. B and the slope C are parameters for each polymer, T is the absolute temperature.

MODEL

Yield behaviour of amorphous polymers has been investigated repeatedly[3-6], and some interesting models have been proposed. Yield stress usually is measured within a comparatively small temperature range adjacent to the glass transition. This is the range in which Robertson's[7] equation for yield stress gives quite useful results. However, Argon[9] points out that Robertson's equation is not applicable below 200 K.

Other theories were introduced by Argon[8-9], Bowden[10] and, more recently, by Kitagawa[11]. They all imply knowledge of the modulus as a function of temperature.

The following approach, based on Eyring's explanation of viscosity[12], simplified some effects (e.g. β-relaxation) but has two advantages: It works and it is simple!

The yield process in a polymer resembles the flow of a liquid. Microscopically, yield may be envisaged as molecular segments giving way step by step under the applied load (Figure 9). The mobility of these segments is determined by their thermal energy. However, there are barriers which the segments must surmount in order to move from one site to another.

The following equation for the shear rate ($\overset{0}{\varepsilon}$) is derived by accepting Eyring's argument[12] and applying it to a yield process[13]

$$\frac{d\varepsilon}{dt} = \overset{0}{\varepsilon} = \frac{P}{2} \exp. \frac{-U + VGy}{kT} \tag{2}$$

The activation energy is U and the activation volume is V (Figure 10). Gy is the yield stress and P a factor that will ultimately be eliminated. The yield stress is arrived at by rewriting the equation (2):

$$Gy = \frac{U}{V} - T\frac{k}{V} \cdot \ln \frac{P}{2\overset{0}{\varepsilon}} \tag{3}$$

This equation for yield stress is linear with temperature (T) and coincides with the results of the experiments carried out

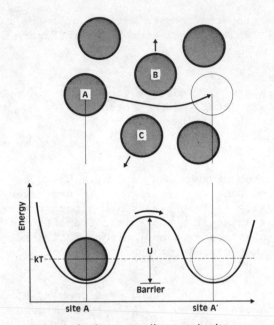

Segment A has to overcome the energy barrier
caused by the segments B and C

Figure 9.   Cross section of a few Molecular Chains directed
            Perpendicular to the Plane

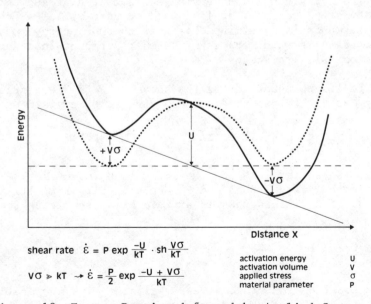

shear rate $\dot{\varepsilon} = P \exp \dfrac{-U}{kT} \cdot \text{sh} \dfrac{V\sigma}{kT}$

$V\sigma \gg kT \rightarrow \dot{\varepsilon} = \dfrac{P}{2} \exp \dfrac{-U + V\sigma}{kT}$

activation energy   U
activation volume   V
applied stress   σ
material parameter   P

Figure 10. Energy Barrier deformed by Applied Stress σ

$$TKF = B - T \cdot C \tag{1}$$

Comparison of the two coefficients leads to the conclusion that

$$B = \frac{U}{V} \tag{4}$$

$$C = \frac{k}{V} \ln \frac{P}{2\varrho} = \frac{U}{V} \frac{1}{T_g} + \frac{k}{V} \ln \frac{kT_g}{2V\mathring{e}\gamma(T_g)} \tag{5}$$

P in equation (5) is eliminated using the viscosity (γeta T_g) at the glass transition-temperature T_g.
Therefore, slope C is dependent on the T_g of the polymer and on the testing speed as well as on activation volume and activation energy. The cohesive strength at absolute zero (B) equals the ratio of activation energy to activation volume. According to the model, B is dependent neither on testing speed nor on T_g. Moreover, experiments (see p4) show that B is independent of crosslink density. Apparently, the cohesive strength of amorphous polymers is primarily due to Van der Waals forces acting between the molecular segments. The force required to break a carbon-to-carbon bond has no bearing on cohesive strength, neither in thermoplasts nor in crosslinked polymers.

Chemical Structure

Six examples serve to illustrate the correlation between chemistry and cohesive strength (Figure 11). B denotes cohesive strength extrapolated to absolute zero, C is the slope, and T_g is the glass transition temperature.

The uncrosslinked phenoxy resin shows a value for B similar to that of crosslinked resins such as diglycidylether of bisphenol-A cured either with diamino-diphenylmethane or with dicyan-diamide. This proves that B is independent of crosslinking density. B = 250 MNm^{-2} is a remarkable but by no means an impossible figure: a flexural strength of 260 MNm^{-2} has been reported[14] for the epoxy resin at temperature of liquid helium (i.e. 4 K).

It is to be expected that activation energy increases with the polarity of the segments. Consequently, it is no surprise to encounter a higher value of B for a resin cured with triethylene tetramine than for one cured with trimethylhexamethylene diamine. Of course, the value for B is also higher if more polar resins are used, e.g. hydantoins.

At temperatures well above asbsolute zero, at room temperature for instance, cohesive strength is also influenced by the slope C. As a rule, C was found to be small if T_g was high and B was low. Crosslinking and rigid segments limit the mobility of the molecules. Hence T_g goes up and C drops. In addition, C is dependent on the

Structure:		Tg °C	$B = \frac{U}{V}$ MN·m^{-2}	slope C MN·m^{-2}·K^{-1}
1	Phenoxy: 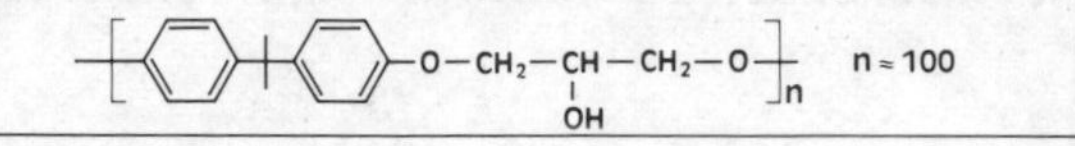 $-[\text{(structure)}-O-CH_2-CH(OH)-CH_2-O-]_n$ n ≈ 100	90	248	0,58
2	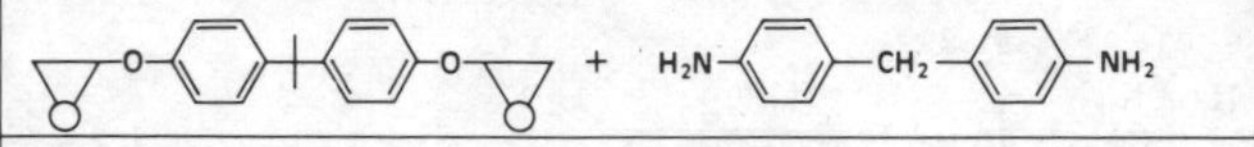 (epoxy structure) + $H_2N-\text{(ring)}-CH_2-\text{(ring)}-NH_2$	159	250	0,54
3	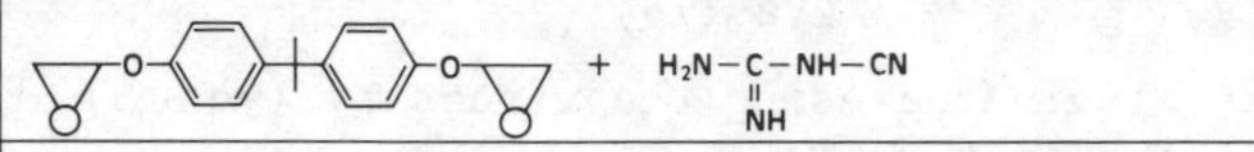 (epoxy structure) + $H_2N-C(=NH)-NH-CN$	120	251	0,52
4	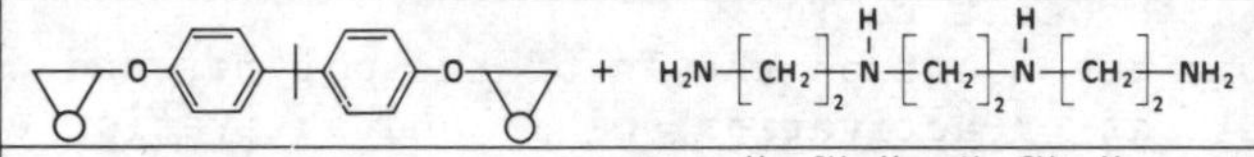 (epoxy structure) + $H_2N-[CH_2]_2-N(H)-[CH_2]_2-N(H)-[CH_2]_2-NH_2$	125	271	0,61
5	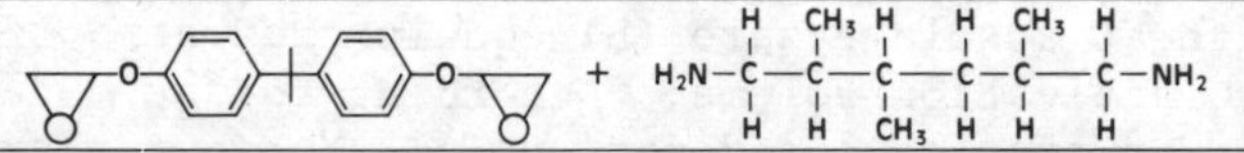 (epoxy structure) + $H_2N-CH_2-CH(CH_3)-CH_2-CH_2-CH(CH_3)-CH_2-NH_2$	100	195	0,47
6	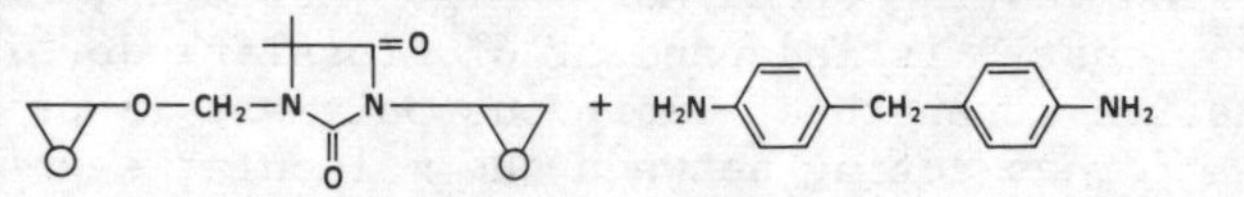 (epoxy structure with hydantoin) + $H_2N-\text{(ring)}-CH_2-\text{(ring)}-NH_2$	177	284	0,58

Figure 11. Correlation of Chemistry and Cohesive Strength (CS); (CS) = B – C · T

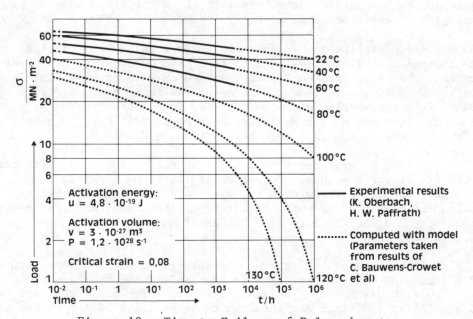

Figure 12. Time to Failure of Polycarbonate

testing speed. All these observations are in accordance with the
model.

Long Term Performance

Assuming there is a critical strain, the model allows the
prediction of the long-term behaviour of polymers versus temperature.
A set of time-to-failure curves was computed for several temperatures
using activation energy and activation volume of polycarbonate
taken from measurements by Bauwens-Crowet et al[15]. The results are
plotted in Figure 12. This shows for instance that a sample
subjected to a load of 20 MNm^{-2} will fail after 2 hours at $130^{o}C$, but
only after 1000 h at $100^{o}C$.

Oberbach and Paffrath[16] measured the time-to-failure of polycarbonate at $22^{o}C$, $40^{o}C$, $60^{o}C$ and $80^{o}C$ over service periods up to
5000 h. Their results are shown as a solid line in Figure 12.
The model offered does not take into account important factors such
as penetration by liquids or moisture and thermal or mechanical
ageing. All these influences depend on the end-use of the applied
polymer. On the other hand, the model enables general long-term-
stability to be predicted to a first approximation using simple
yield experiments. At low temperatures, excellent long-term-stability
and creep resistance are to be expected and even polymers with low
polarity and low yield strength at room temperature are strong
enough at low temperatures and so may be considered for cryogenic
applications.

Flexibilization

Formulated products often incorporate a flexibilizer such as
Hycar - CTBN or aliphatic polyesters. Even the addition of small
amounts of flexibilizer e.g. 15% by weight will markedly improve
fracture toughness, impact strength and peel strength. However,
the improvement of some properties may impair others:shear strength
and therefore cohesive strength are both reduced by flexibilizers.

Figure 13 shows shear strength (in torsion) versus temperature
for the original unflexibilized resin and for three flexibilized
versions. Flexibilization results in a reduction of slope C and a
loss of cohesive strength of up to 25%. This loss is unimportant,
as the cohesive strength remains high enough to suffice for low
temperature applications.

A polymer characterized by a low cohesive strength versus
temperature gradient is required when low temperature strength is
to be paired with sufficient room temperature strength, i.e. a
polymer with a small C value must be sought. Such a polymer may be
arrived at by combining flexible segments of low polarity (e.g.
polymethylene segments) with rigid and bulky segments or crystalline
regions.

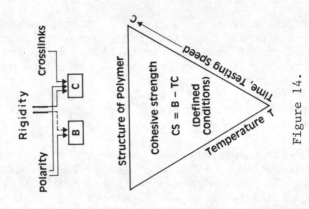

Figure 14.

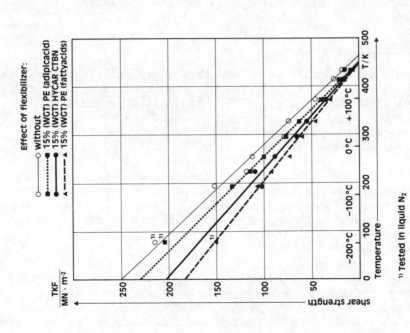

Figure 13. Shear Strength versus Temperature
AL bonded to AL with DGEBA + MDA

CONCLUSIONS

Cohesive strength is a function of three variables:
Polymer structure, time and temperature (Figure 14). The model
described depicts the relationship between cohesive strength, polymer
structure, and time over a wide range of temperatures. Such a
relationship is particularly valuable for low temperature
applications where polymers are occasionally exposed to room
temperature as well.

ACKNOWLEDGEMENTS

I am grateful to the head of the research department, Prof.
Dr. H Batzer and to Dr. H. Gysling for the continuous support they
have given me in my work.
Dr R Schmid has initiated research in cohesive strength of polymers.
I greatly appreciate his contributions and his helpful advice,
originating from his prfound knowledge of those materials. I thank
Mr R Lopez and Mr W Kessler who have carried out many of the
experiments described and Mr S A Napier who was kind enough to read
the manuscript.

REFERENCES

1. A. Kelly, "Strong Solids", 2nd edition, Clarendon Press,
   Oxford, 1973 p.12

2. Modern Plastic Encyclopedia 1979-1980 p.511

3. I. M. Ward, "Review: The Yield Behaviour of Polymers", J.
   Material Sci. 6 (1971), 1397-1417

4. S. S. Sternstein, "Yielding Modes in Glassy Polymers" in
   "Polymeric Materials", ed. E. Baer, American Society for
   Metals, Metals Park, OHIO, 1975

5. P. B. Bowden "The Yield Behaviour of Glassy Polymers" in
   "The Physics of Glassy Polymers" ed. R. N. Haward, Applied
   Science Publishers Ltd, London, 1973.

6. S. Yamini, R. J. Young, "The Mechanical Properties of Epoxy
   Resins", J. Material Sci. 15 (1980), 1814-1822.

7. E. R. Robertson, "An Equation for the Yield Stress of a
   Glassy Polymer", Appl. Polym. Symposia 7 (1968), 201-213.

8. A. S. Argon, "A Theory for the Low Temperature Plastic
   Deformation of Glassy Polymers" Phil. Mag. 28 (1973), 839-865.

9. A.S. Argon, M.J. Bessonov, "Plastic Deformation in Polyimides, with new implications on the Theory of Plastic Deformation of Glassy Polymers", Phil. Mag. 35 (1977), 917-933.

10. P. B. Bowden and S. Raha, "A Molecular Model for Yield and Flow in Amorphous Glassy Polymers making use of a Dislocation Analogue", Phil. Mag. 29 (1974) 149-166.

11. M. Kitagawa, "Power Law Relationship Between Yield Stress and Shear Modulus for Glassy Polymers", J. Polym. Sci. Phys. Ed. 15 (1977), 1601-1611.

12. H. Eyring, "Viscosity, Plasticity and Diffusion as Examples of Absolute Reaction Rates", J. Chem. Phys. 4 (1936), 283-291.

13. M. Fischer, F. Lohse, R. Schmid, "Struktureller Aufbau und Physikalisches Verhalten Vernetzter Epoxidharze" Markomol. Chem. 181 (1980), 1251-1287.

14. D. Evans, J. T. Morgan, G. B. Stapleton, "Epoxy Resins for Superconducting Magnet Encapsulation", Rutherford Laboratory Report SNB 90 2376454 (1972).

15. C. Bauwens-Crowet, J. C. Bauwens, G. Homes, "Tensile Yield - Stress Behaviour of Glassy Polymers", J. Polym. Sci. Part A-2, 7 (1969), 735-742.

16. K. Oberbach, H. W. Paffrath, "Dehn- und Festigkeitsverhalten von Kunststoffen im Zeitstand Zugversuch", Materialprufung 4, (1962) 291-296.

THE THERMAL CONDUCTIVITY AND THERMAL EXPANSION OF

NON-METALLIC COMPOSITE MATERIALS AT LOW TEMPERATURES

H M Rosenberg

The Clarendon Laboratory
University of Oxford
Oxford OX1 3PU, UK

INTRODUCTION

This paper presents a short review of the behaviour of the thermal conductivity and the thermal expansion of non-metallic particle-filled and fibre-filled composite materials from room temperature down to liquid helium temperatures. A brief discussion of the physical mechanisms which are involved and the possibility of predicting the thermal properties of a composite is also given.

THE THERMAL CONDUCTIVITY OF THE FILLER MATERIAL

The heat transport in non-metals is determined by the flow of lattice vibrational energy (phonons) along the specimen. This phonon flow can be considered to be similar to the propagation of high frequency sound waves. The phonons will be scattered by various types of imperfections in the material and it is this scattering which limits the heat flow and gives rise to a thermal resistance. As the temperature is reduced from room temperature down to liquid helium temperatures the dominant phonon wavelength increases from being a few atomic spacings (say 0.5 to 1 nm) to many hundreds or thousands of atomic spacings. In order for scattering to be significant the size of the scattering centres must be of the order of the phonon wavelength and so, whilst at high temperatures the thermal resistance can be limited by the presence of impurity atoms (and also by the scattering of one phonon by another in very pure materials) at low temperatures only large scale defects, such as the size of the individual crystallites that make up the material are able to limit the conductivity. In single crystals the ultimate limitation is that of the boundaries of the specimen because the phonons are able to travel unimpeded

right across the sample and they are only scattered at its surfaces.
The phonon mean free path, 1, is given by the smallest dimension of
the sample. The thermal conductivity, κ, can then be calculated by
using the standard kinetic theory expression

$$\kappa = \frac{1}{3} c \, v \, 1 \qquad\qquad\qquad\qquad\qquad (1)$$

where c is the specific heat per unit volume, v is the velocity of
sound and 1 is the mean free path of the heat carriers. This
expression gives a very good estimate of the thermal conductivity
for single crystals at low temperatures, if 1 is set equal to the
smallest dimension of the specimen. A full discussion and review
is given in the book by Berman.[1]

     Equation (1) implies that the conductivity of a particle or
of a fibre at low temperatures will not be that of the bulk material
but it will be reduced to a value which is proportional to the mean
diameter of the particle or fibre. It will also have the same
temperature variation as the specific heat (ie T^3) since all the
other quantities in equation (1) are constant.

THE THERMAL CONDUCTIVITY OF THE MATRIX

     The most commonly used matrices for the composites which we
shall consider are epoxy-resins. These are glassy materials and
due to their very irregular atomic structure the phonon scattering
is very strong, the mean free path is small and so they have a low
conductivity. They also all have a rather unusual but characteristic
feature - there is a plateau in the thermal conductivity in the
range 5-10 K (eg see Fig 1(a)). At temperatures below this plateau
the conductivity drops off with an approximately $\underline{T}^2$ dependence.
It has a similar magnitude for all resins.

ACOUSTIC MISMATCH

     The difficulty of predicting the thermal conductivity from the
properties of the individual constituents lies in the fact that
there is a further resistive mechanism which occurs in a two-phase
system. When the phonons pass from matrix to filler and vice
versa there is additional scattering at the interface due to
acoustic mismatch. This is a phenomenon which arises whenever
energy is transferred from one medium to another and for optimum
transmission the relevant impedances of the two media must be
equal. For heat conductivity at low temperatures it is the
acoustic impedance (density x sound velocity) which controls the
mismatch and because, in general, the density and sound velocity
in epoxy-resin are both rather low compared with most filler
materials this leads to considerable extra scattering when almost
any crystalline filler is used. The effect of the mismatch will
be increased if, for a given volume conventration, V_f, small filler

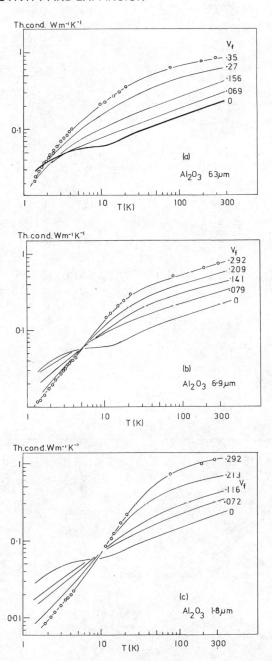

Fig 1.   The thermal conductivity of epoxy/Al$_2$O$_3$ (corundum) composites
         for three sizes of filler. Note that the decrease in
         conductivity in the helium range, which is due to acoustic
         mismatch, is greatest for the smallest particles.[2]
         (a) 63 μm   (b) 6.9 μm   (c) 1.8 μm.

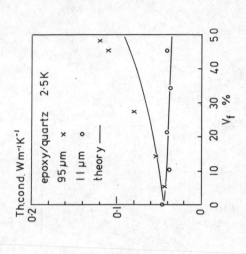

Fig. 3. The thermal conductivity of epoxy/quartz composites at 2.5 K as a function of V_f for 11 and 95 μm particles.[2] The solid lines are from calculations[7] based on the theory of Hamilton and Crosser.

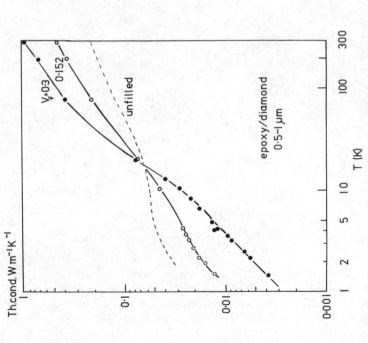

Fig. 2. The thermal conductivity of epoxy/diamond composites (½ – 1 μm) showing the very strong acoustic mismatch effect at liquid helium temperatures (R A Boone).

particles are used, because then the effective surface area at which the mismatch occurs will be increased. The mismatch effects are illustrated by the results for an epoxy/Al_2O_3 (corundum) composite[2] in Fig 1. Fig 1(a) (65 μm particles) shows the general way in which the thermal conductivity at high temperatures is increased by the addition of filler whereas at liquid helium temperatures it is depressed below that of pure epoxy resin due to acoustic mismatch. It should be noted that the thermal conductivity of an Al_2O_3 single crystal is usually very high in this region and so the mismatch effect exerts a very large influence on the conductivity. Fig 1(b) (6.9 μm) and (c) (1.8 μm) show how the scattering becomes more pronounced when smaller filler particles are used.

The most extreme mismatch we have found is for an epoxy/ diamond composite system. Diamond has a relatively high density ($\sim$ 3500 kg m^{-3}) and a very high sound velocity ($\sim$ 10^4 m s^{-1}). Some results are shown in Fig 2 where it will be seen that the addition of 30 volume percent $\frac{1}{2}$ - 1 μm diamond powder decreases the thermal conductivity of the resin by a factor of about 10 at 2 K.

The smallest mismatch is probably given by glass and certainly the thermal conductivity of epoxy/ballotini does not show the effect to any large extent.[2]

Although not strictly relevant to this paper it should be mentioned that composites with metal fillers also show the mismatch effect[3] although generally it is not so pronounced.

THERMAL CONDUCTIVITY OF PARTICLE-FILLED COMPOSITES

Both Maxwell[4] and Lord Rayleigh[5] developed theories for the conductivity of an array of spherical particles in a matrix and many semi-empirical expressions have also been proposed. We have found that the basic formula of Rayleigh, which treats a system of interacting spheres in a matrix, to be a satisfactory basis for interpreting many of our results. This formula has been extended to higher powers by Meredith and Tobias[6] which is necessary for high volume concentrations of filler and this is the form which we have used.[2]

If the conductivity of the filler is much higher than that of the matrix (say 100 times) then the shape of the particle can become important. The parameter which is used to designate the shape is the sphericity, ψ. This is the ratio of the surface area of a sphere with the same volume as that of the particle to the surface area of the particle. The sphericity has been introduced into a modification of Maxwell's original formula by Hamilton and Crosser.[7] It has the disadvantage of the original Maxwell theory in that it does not take account of the mutual interaction of the particles and so it is not so satisfactory a treatment for high V_f. Details of these formulae are given in reference 2.

None of these expressions, however, take account of acoustic mismatch and so whilst they give reasonable agreement with experiment at high temperatures they must be modified for use in the liquid helium range. One way of doing this is to estimate the contact resistance between particle and matrix and to add this to the ordinary thermal resistance of a particle. This then yields an 'effective' conductivity of the particle which can be used in the standard formulae. The type of agreement which can be achieved with experiment using the Hamilton and Crosser expression is illustrated for epoxy/quartz composites containing 95 and 11 µm particles in Fig 3.

If the acoustic mismatch resistance is very large compared with the resistance of both filler and matrix good agreemement cannot be obtained and a further expression has been developed[2] which must be used.

THE THERMAL CONDUCTIVITY OF FIBRE-FILLED COMPOSITES

We limit our discussion here to composites in which all fibres are parallel to one another. Very different results are obtained depending on whether the heat flow is parallel or perpendicular to the fibre direction.

In the parallel direction the system can be considered to consist of two conducting paths in parallel with very little interaction between them. For this situation the simple formula for mixtures

$$\kappa = V_f \kappa_f + (1-V_f)\kappa_m \qquad\qquad\qquad (2)$$

is reasonably satisfactory (where κ_f and κ_m are the conductivity of filler and matrix respectively). κ_f can be found by measuring the conductivity of a bundle of fibres.

For epoxy/glass-fibre composites (GFRP) agreement between theory and experiment to within 5% over the range 1-80 K has been achieved for V_f from 0.33 to 0.77,[8] see Fig 4.

For epoxy/carbon (graphite) fibre composites (CFRP) the agreement is not so good below about 8 K. This is due to the fact that the acoustic mismatch is much more serious for epoxy/carbon than for epoxy/glass and since, of course, the phonons in the resin do not travel exactly parallel to the fibres this will have the effect of reducing the phonon mean free path in the epoxy below its bulk value[8] (see Fig 5).

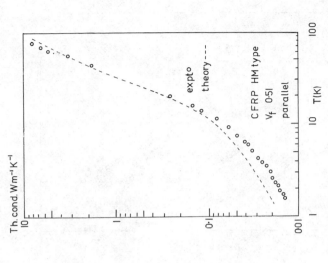

Fig. 5. The thermal conductivity of HM type CFRP ($V_f = 0.51$) in a direction parallel to the fibres compared with calculations using equation (2).

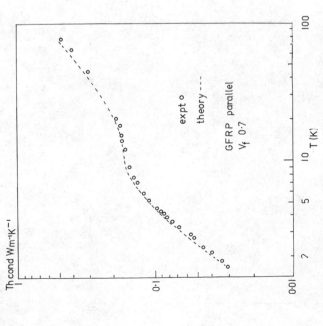

Fig. 4. The thermal conductivity of GFRP ($V_f = 0.7$) in a direction parallel to the fibres compared with calculations using equation (2).

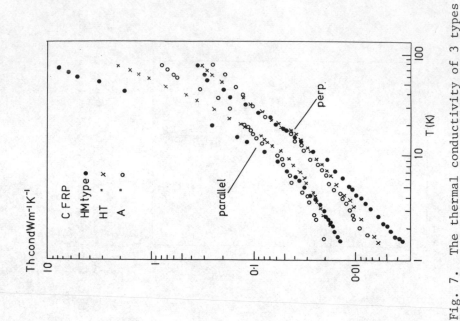

Fig. 7. The thermal conductivity of 3 types of Courtaulds Grafil CFRP in directions parallel and perpendicular to the fibres.[8],[15]

Fig. 6. The thermal conductivity of GFRP ($V_f = 0.77$) in a direction perpendicular to the fibres[5] compared with calculations using Rayleigh's[5] theory. The drop in conductivity below the theoretical prediction is because[8] acoustical mismatch has not been included.[8]

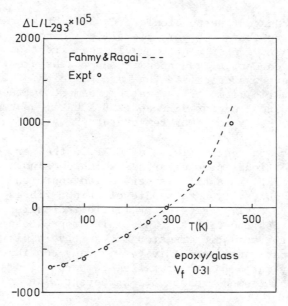

Fig 8. The thermal expansion of epoxy/glass ballotini ($V_f = 0.31$) compared with calculations from the theory of Fahmy and Ragai.[11-13]

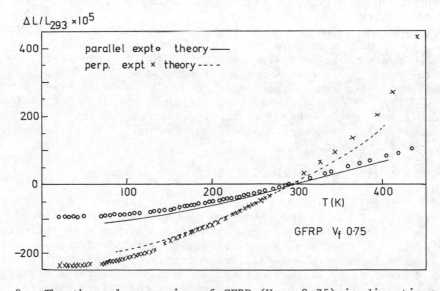

Fig 9. The thermal expansion of GFRP ($V_f = 0.75$) in directions both parallel and perpendicular to the fibres. The calculations are based on the theories of turner[17] and Schapery[18] (parallel) and Chamberlain[19] (hexagonal array, quoted in reference 20) (perpendicular).[14,15]

The heat transport perpendicular to the fibres presents a similar problem to that of a particle-filled composite because one of the phases is no longer continuous in the direction of the heat flow. The heat conductivity perpendicular to the axes of a cubic array of cylinders was calculated by Lord Rayleigh.[5] Once again because the acoustic mismatch is small agreement between theory and experiment is quite good for GFRP above 8 K but measured values are 50% down on the theory at lower temperatures[8] (Fig 6).

To calculate the conductivity of CFRP in the perpendicular direction an additional problem arises which does not need to be considered for GFRP. Graphite is very anisotropic and hence the longitudinal conductivity of a bundle of carbon fibres (which can be measured quite easily) gives very little indication of the conductivity across the fibres. This anisotropy, coupled with the uncertainty in the acoustic mismatch makes it very difficult to calculate the transverse conductivity with any confidence although if an anisotropy ratio for the parallel to perpendicular conductivities of 0.1 is assumed, reasonable agreement with measured values can be achieved.[8]

Experimental results for three types of Courtaulds Grafil pultruded CFRP are shown in Fig 7.

THE THERMAL CONDUCTIVITY OF CARBON-CARBON COMPOSITES

Composites in which carbon fibres are incorporated in a pyrolitic carbon matrix have recently been investigated.[9] These have an exceedingly small thermal conductivity, which seems to be lower than that of any other constructional material. At 4 K the conductivity of a sample incorporating a 90° woven cloth was about 8×10^{-3} W m^{-1} K^{-1} in a direction parallel to one set of fibres and perpendicular to the other: in the direction perpendicular to both sets of fibres it was about half of this value. Further details are given in reference 9.

THE THERMAL EXPANSION OF COMPOSITES

The thermal expansion of a solid is controlled by the extent to which the motion of the vibrating atoms deviates from simple harmonic motion (SHM). If the atomic displacements were strictly simple harmonic then the thermal expansion would be zero. As the temperature is reduced any deviations from SHM or anharmonicities as they are called, become very small and hence the thermal expansion (or contraction) is reduced at low temperatures. In fact below about 50 K it is a good approximation to assume that the expansion coefficient of any constructional material is zero.

The calculation of the thermal expansion of a composite from the properties of its constituents is complicated by the fact that,

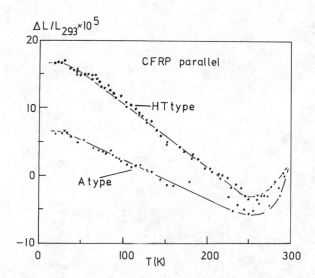

Fig 10.   The thermal expansion of two types of Courtaulds Grafil
          CFRP in a direction parallel to the fibres. Note that
          the material expands slightly on cooling.[14,15]

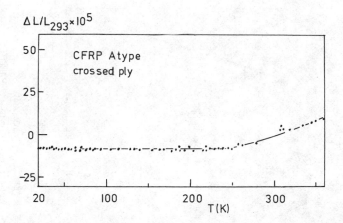

Fig 11.   The thermal expansion of a 90° crossed-ply CFRP made from
          Courtaulds Grafil A type fibre which shows virtually no
          dimensional change below 250 K.[15,16]

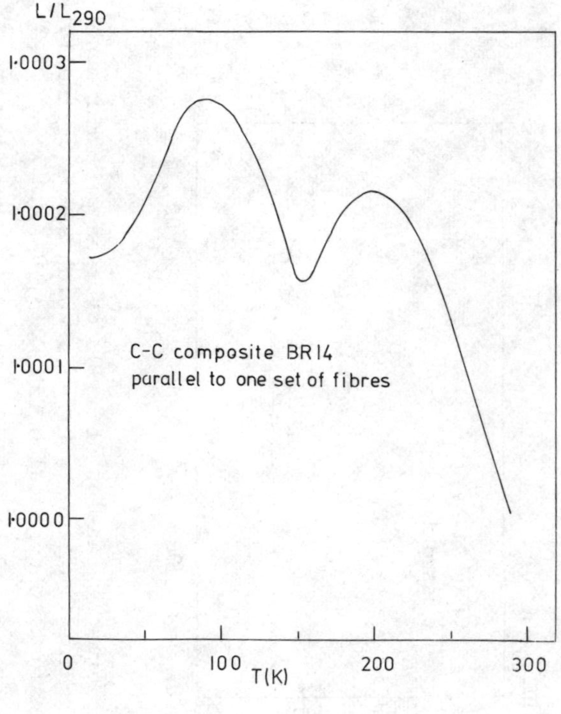

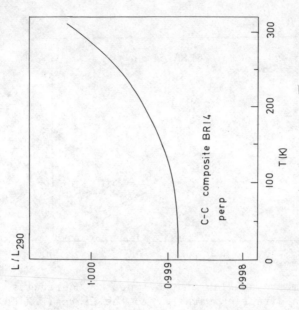

Fig 13. The thermal expansion of a 90° cross-woven carbon-carbon composite, BR14 (V_f = 0.45) in a direction perpendicular to both sets of fibres.

Fig. 12. The thermal expansion of a 90° cross-woven carbon-carbon composite BR14 (V_f = 0.45) in a direction parallel to one set of fibres and perpendicular to the other set.

due to the differential expansion between matrix and filler, stresses
are set up at the interfaces and so the calculation of the final
effective dimensional change of the specimen is a function not only
of the individual expansivities but also of the elastic constants.
Very many formulae have been proposed and we give no details here –
a review is given by Holliday and Robinson.[10]

For composites with an epoxy-resin matrix – which like all
polymers has a very high expansion coefficient at room temperature –
the general effect of adding either a particle or a fibre filler is
to reduce the expansion.

THE THERMAL EXPANSION OF POWDER-FILLED COMPOSITES

A formula which seems to give reasonable agreement with experiment is that proposed by Fahmy and Ragai[11] who assumed that the
particles were spherical and that the regions of stress caused by
individual particles did not overlap with one another. An example
is shown in Fig 8 for an epoxy/ballotini composite with $V_f = 0.31$.
While this shows good agreement between the measured and calculated
values, measurements on epoxy/quartz and epoxy/Al_2O_3 do not show
such satisfactory agreement with theory because (a) the particles
are not spheres and (b) there are greater differences between the
expansion coefficients and the elastic constants of matrix and
filler.[12,13]

THE THERMAL EXPANSION OF FIBRE-FILLED COMPOSITES

In general the expansion in the direction parallel to the
fibres is considerably smaller than in the perpendicular direction
because of the greater area of surface contact in the parallel
direction – which will inhibit the expansion of the matrix.
Formulae similar to those used for particle filled composites can
be used to calculate the expansion and these yield reasonable
agreement with experimental values.[14,15] Results for a typical
GFRP are shown in Fig 9.

The thermal expansion of CFRP is especially interesting because
graphite has a negative coefficient of thermal expansion parallel
to the basal plane and since the axis of carbon fibres lies in that
plane this means that they expand along their length on cooling and
contract on heating. The dimensional change of the fibres is therefore directly counter to that of the resin matrix and hence the
expansion of CFRP tends to be exceedingly small. Depending on the
fibre material and V_f the expansion coefficient can be either
negative or positive, but in general it is extremely low at room
temperature and it is negative as the temperature is reduced.
Results for two types of Courtaulds Grafil pultruded CFRP are
shown in Fig 10.[14]

Perpendicular to the fibre axis, the thermal expansion coeffi-
cent is positive and is not particularly small and hence in that
direction the dimensional change of CFRP is similar to that of
other materials. However this does lead to the possibility that a
suitable combination of carbon fibre plies at 90° to one another
can give a material in which the positive effect parallel to one
set of plies on cooling is just cancelled by the negative effect
due to the perpendicular contraction of the other set. An example
of such a CFRP material is shown in Fig 11 where it will be seen
that below about 250 K there is no detectable dimensional change.[15,16]

Carbon-carbon composites also show rather complicated behaviour
on cooling but once again the dimensional changes are small. For
one 90° cross-woven sample in the direction perpendicular to one
set of fibres and parallel to the others there is an overall length
increase of about 2 parts in 10^4 on cooling to 4 K (Fig 12).
Perpendicular to both sets of fibres there is an overall contraction
of about 1 part in 10^3 over the same temperature range (Fig 13).
This low thermal expansion coupled with the extremely small thermal
conductivity mentioned in a previous section would seem to make
carbon-carbon composites worthwhile considering as a constructional
material in special circumstances when high dimensional stability
and very good thermal insulation are required, although in many
cases CFRP might be equally satisfactory.

ACKNOWLEDGEMENT

This review has drawn heavily on the work done by several of
my research students to whom all credit is really due.

REFERENCES

1.   R Berman, Thermal Conduction in Solids, Oxford University
        Press (1976).
2.   K W Garrett and H M Rosenberg, J. Phys. D: Appl. Phys. 7,
        1247 (1974).
3.   F F T de Araujo and H M Rosenberg, J. Phys. D. Appl. Phys. 9,
        665 (1976).
4.   J C Maxwell, A Treatise on Electricity and Magnetism, Vol. 1
        (3rd edn.) pp. 435-41, Oxford, Clarendon Press (1892).
5.   Lord Rayleigh, Phil. Mag. 34, 481 (1892).
6.   R E Meredith and C W Tobias, J. Appl. Phys. 31, 1270 (1960).
7.   R L Hamilton and O K Crosser, Ind. Engng. Chem. Fundamentals
        1, 187 (1962).
8.   D J Radcliffe, D. Phil. thesis, Oxford University (1979).
9.   C I Nicholls and H M Rosenberg, Proc. I.C.E.C. 8, Genova,
        p. 715, I.P.C. Ltd (1980).
10. L Holliday and J Robinson, J. Mat. Sci. 8, 301 (1973).
11. A A Fahmy and A N Ragai, J. Appl. Phys. 41, 5108 (1970).
12. J A Grubb, D. Phil. thesis, Oxford University (1975).

13.   J A Grubb and H M Rosenberg, Proc. Int. Conf. on Composite
        Materials, p. 602 (1975).
14.   M de F F Pinheiro, D. Phil. thesis, Oxford University (1978).
15.   M de F F Pinheiro, D J Radcliffe and H M Rosenberg, Proc.
        I.C.E.C. 7, London, p. 494, I.P.C. Ltd (1978).
16.   M de F F Pinheiro and H M Rosenberg, Cryogenics 18, 373 (1978).
17.   P S Turner, J. Res. N.B.S. 37, 239 (1946).
18.   R A Schapery, J. Comp. Mat. 2, 38 (1968).
19.   N J Chamberlain, BAC Rep. No. SON(P) 33 (1968).
20.   K F Rogers et al, J. Mat. Sci. 12, 718 (1977).

NONLINEAR STRESSES AND DISPLACEMENTS OF THE FIBRES AND

MATRIX IN A RADIALLY LOADED CIRCULAR COMPOSITE RING +

Samaan George Ladkany *

The Johns Hopkins University, Baltimore, MD
and
The University of Wisconsin, Madison, WI

INTRODUCTION

The mechanical behaviour and failure mechanism of the
reinforcing fibres and the polymer matrix of a circular composite
ring, subjected to radial pressure and cryogenic thermal loading,
and the design of a thick composite arch, are studied using nonlinear finite element analysis. The rings are sections of a
pultruded 0.203 m x 0.0065 m (8 in x $\frac{1}{4}$ in) diameter x thickness,
polyester-glass pipe, reinforced in both the axial and circumferential directions. The arch is 1.0 m in radius, 0.1 m thick and
0.065 m wide. Circular rings of pultruded polyester-glass may be
used at the tips of the composite radial structures, supporting the
round aluminum rippled conductors of superconductive energy storage
magnets.

We present results of preliminary analytical studies of polyester-glass rings having complete or partial circumferential glass
reinforcements and investigate the effect of cavities and crack
propagation, in the matrix, on the distribution and concentration
of the stresses in the reinforcing circumferential glass strands
and in the polyester matrix.

+Superconductive Energy Storage Project, University of Wisconsin-
 Madison.

*At present, Assistant Professor, Department of Civil Engineering/
 Mat. Sci. & Eng., The Johns Hopkins University.

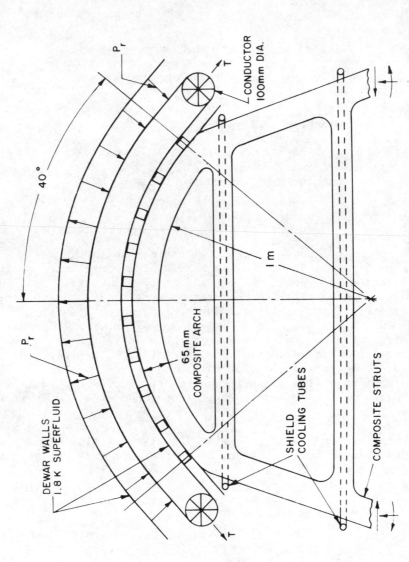

Figure 1. Glass composite strut and supporting arch for an energy storage manget

FIBREGLASS STRUCTURAL COMPOSITES FOR ENERGY STORAGE MAGNETS

Energy storage magnets[1] are in the form of large solenoids, constructed in underground tunnels, and supported by fibreglass composite structures that transmit the magnetic forces from the 1.8 K conductors to the room temperature bedrock. The magnets are large potential users of fibreglass-polymer composites; a single 1000 MWh unit would require 1.6 Gg of material. Fibreglass composites are ideal as structural supports for the superconductive storage magnets due to their high strength and low thermal conductivity. The large mass of composite materials needed for the structural supports (struts) in the magnet necessitates the use of commercially fabricated, easily obtainable, low-cost composites. Test results of various industrial laminates[2,3] at 300 K, 77 K, and 4.2 K indicated that unidirectional glass-polyester had high strengths, especially in compression, comparable with glass-epoxies, but was one-half to one-third as expensive. Several structural designs of the laminated fibreglass composites which support the Energy Storage Magnets have been investigated.[4] The analysis indicated the need for composite structural shapes such as pipes, I-beams, thick plates and T-sections. Such structural shapes are commercially available in bulk quantities and at low costs in pultruded glass-reinforced polyester. The round aluminum conductors[5] in the magnets are rippled and supported at intervals of 3-8 m, by circular composite rings at the tip of the struts (Fig 1). The entire conductor-strut assembly is subjected to cyclical loading once a day during the fifty year life of the magnet. The composite struts are cryogenically shielded at 11 K and 77 K and are fastened to the tunnel walls via metal base plates and rock bolts.

PULTRUDED GLASS REINFORCED POLYESTER

The rings and structures investigated are constructed of pultruded glass reinforced polyester composites. An example of the material used is EXTREN, manufactured by the Morrisson Fibre-glass Manufacturing Corporation. Chopped glass fibres of various lengths are randomly mixed with the polyester. Glass strands, manufactured by Owens Corning, are added unidirectionally in some structural shapes such as plates, and bidirectionally in others, such as pipes. Longitudinal glass reinforcements are required for the process of pultrusion in which the composite is pulled through a metal die. The total glass content varies from 44% by weight for unidirectional plates to over 55% by weight for pipes. It is very important to note that the pultrusion process results in a nonuniform distribution of the glass strands; the strands vary in thickness up to 0.5 mm, are placed 1 mm to 5 mm apart in the longitudinal direction, and are sparsely spaced up to 25 mm, in layers carrying crosswise circumferential reinforcements in pipes. The ratio of lengthwise to crosswise reinforcement is approximately

TABLE 1:   PROPERTIES OF PULTRUDED GLASS POLYESTER (EXTREN)
           AT 300 K AND 77 K

PROPERTY/TEMP	300 K		77 K	
	FIBRE DIRECTION	CROSSWISE	FIBRE DIRECTION	CROSSWISE
TENSILE STRENGTH	138 MPa	69 MPa	207 MPa+	90 MPa+
COMPRESSIVE STRENGTH	240 MPa*	200 MPa*	359 MPa	262 MPa
TENSILE MODULUS	13.8 GPa	10.4 GPa	16.6 GPa*	12.4 GPa*
FLEXURAL STRENGTH	241 MPa*	124 MPa		
BOND STRENGTH	16.5 MPa*	18.9 MPa*	20 MPa+	24 MPa+
SHEAR STRENGTH	41.5 MPa		54 MPa+	
SPECIFIC GRAVITY	1.74 gm/cc			

EXTREN: A PRODUCT OF MORRISON FIBREGLASS MANUFACTURING CORPORATION.

+   PRELIMINARY AND ESTIMATED VALUES.

*   U.W. EXPERIMENTS: YOUNG, KHIM & HAN.[6,7]

5 to 1. The outermost glass layers are protected by 0.5 mm of
treated polymer and the subsequent glass layers are 2 mm apart.
Some of the construction details of an EXTREN plate are shown in
the two photographs of Figs 2 and 3. Various mechanical tests of
EXTREN at 300 K and 77 K have been conducted[6,7] and some of the
results are shown in Table 1.

ANNULAR HYBRID ELEMENTS

     Specialised annular finite elements were derived by the author[8]
for the exact geometrical modelling of circular structures such as
arches, rings, circular beams and plates. These annular elements
are modified to accommodate symmetrically layered circular structures such as the polyester glass composite arches and rings
analysed in this paper; Fig 4 shows the symmetrical three layer
element used. The two outer layers and the middle layer could have
different thicknesses and materials. The principal material axes
of the layers may be independently oriented with respect to the
polar coordinates of the element. A fourth order transformation
tensor is used to obtain the elastic and physical constants of the
layers in the radial, r-direction and the circumferential,

Figure 2. Details of EXTREN matrix.
Complements of K S Han.

Figure 3. EXTREN plate surface showing the glass bands.
Complements of K S Han.

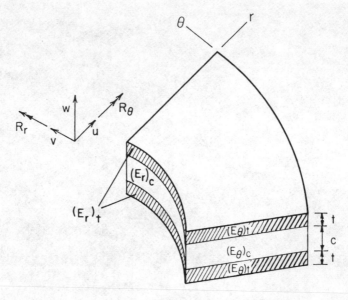

Figure 4.   Details of a layered hybrid element

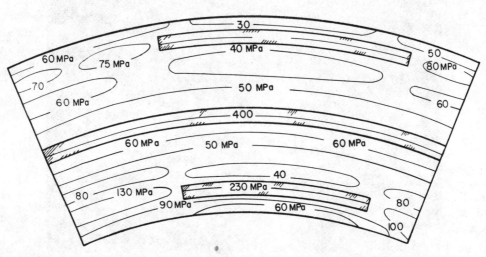

Figure 5.   Stress distribution in a void free ring having a
complete circumferential glass band

θ-direction. The elements are based on the hybrid finite element principle of T. H. H. Pian and Pin Tong[9] for assumed stresses and boundary displacements.

The hybrid functional Π_H used in the derivation of the hybrid finite elements is:

$$\Pi_H = \sum_n \int_{v_n} \frac{1}{2} C_{ijkl} \sigma_{ij} \sigma_{kl} \, dv - \int_{\partial v_n} T_i U_i \, ds + \int_{S_{\sigma_n}} \overline{T}_i U_i \, ds \qquad (1)$$

where:

$$\partial V_n = s_n + s_{\sigma_n} + s_{u_n}$$

is the entire boundary of v_n. S_{σ_n} is the portion of ∂v_n, where the surface tractions $\overline{T}_i$ are specified and S_{u_n} is the portion of ∂V_n, where displacements $\overline{U}$ are specified. That is:

$$U_i = \overline{U}_i \text{ on } S_{u_n}$$

$$T_i = \overline{T}_i \text{ on } S_{\sigma_n}$$

σ_{ij} = Stress tensor

C_{ijkl} = Elastic compliance tensor

V = Volume

∂v = Boundary of V

S = Surface area

ds = Differential element of surface area

T_i = Surface traction (force per unit area)

$\overline{U}_i$ = Prescribed boundary displacements

S_u = Portion of ∂v over which the boundary displacements are prescribed.

The stress tensor, σ_{ij}, must satisfy the equilibrium equations

$$\sigma_{ij,j} + \overline{F}_i = 0, \qquad (2)$$

where $(i,j,k,1 = 1,2,3)$ and $\overline{F}_i$ are the prescribed body forces. The
tensor, σ_{ij}, is also in equilibrium with the prescribed boundary
tractions, $\overline{T}_i$, over the boundary S_σ.

In the hybrid finite element method, the assumed stress field
need not be continuous across the inter-element boundaries, but
equilibrium must be maintained for the surface tractions, T_i =
$\sigma_{ij}\nu_j$, where ν_j are the components of the unit vector normal to the
boundary. In the algorithm, n elements are considered to span the
continuum being analysed with each element sharing nodal points and
boundaries with neighbouring elements. It is worth noting that
advanced finite elements, such as the Isoparametric or Hybrid, use
numerical integration techniques like the Gauss quadrature to
construct their element stiffness matrices. Thus if the radial
width of the element Δr is small compared to its radius r and, if
$\Delta r/r <.02$, numerical computational errors may result.[10] Thus in the
analysis of the composite rings a special topological mapping
technique is used to circumvent the possibility of numerical integration errors.

ANALYSIS OF POLYESTER GLASS RINGS

The composite rings analysed are 0.203 m x 0.0065 m (8 in x
$\frac{1}{4}$ in) and are obtained from sections normal to the axis of symmetry
of EXTREN pipes. Since the glass reinforcement is approximately
five times higher in the axial direction of the pipe than in its
circumferential direction, five ring sections each 10 mm in thickness are considered. Three of these sections contain one circumferentially complete glass band and two partial glass bands each
(Figs 5, 6 and 9). Each of the other two ring sections contain
three partial glass bands (Figs 7 and 8). The partial glass bands
are parts of the long axial reinforcement in the composite pipe.
In the analysis, the composite ring sections are subjected to
internal radial pressure of 6.61 MPa, which results in average
circumferential (hoop) stresses close to the ultimate strength of
the polyester/chopped glass matrix. In Figs 5, 6, 7 and 8 the
stresses in cavity (void) free rings subjected to internal pressure
are shown and compared to rings having various cavities in their
matrices. The cavities are 2 mm to 3 mm long and 0.5 mm to 1.00 mm
wide; they are located at the tip of the glass bands, at their
inner side, or at their outer side. The cavities vary in their
proximity to each other. Stress distribution in the various regions
of the ring sectors analysed are shown. The plots are not contour
lines of the stresses; they show, however, the highest stress or
stress intensity factor in the region enclosed within the curve
indicating the approximate boundary of that region. In Fig 9 the
post failure stress distribution in a ring sector is also shown;
the two sides of the ring are held together after failure by one
continuous circumferential glass band. Annular portions of the
composite rings, 6.5 mm wide and subtended by a central angle of

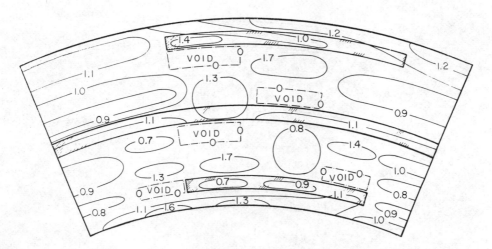

Figure 6.   Stress intensity factors in a ring with voids having
a complete circumferential glass band

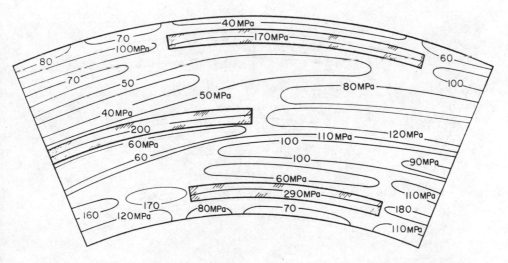

Figure 7.   Stress distribution in a void free ring having
partial circumferential glass bands

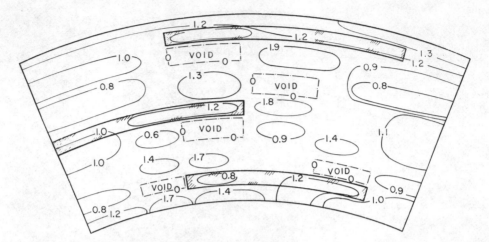

Figure 8.   Stress intensity factors in a ring with voids having
partial circumferential glass bands

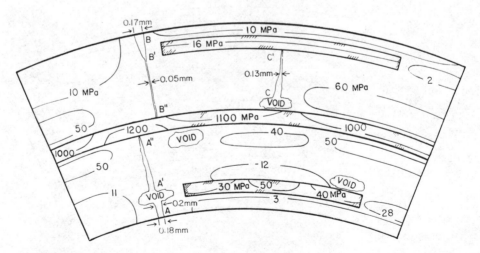

Figure 9.   Stress distribution in a cracked composite ring
having a complete circumferential band

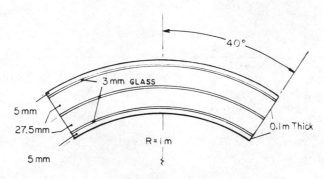

a. ARCH CONSTRUCTION

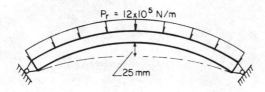

b. PINNED ARCH

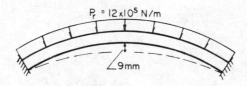

c. FIXED ARCH

Figure 10. Composite arch for an energy storage magnet to be used for carrying the round conductors

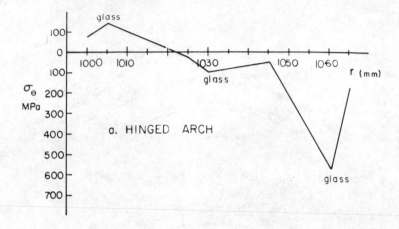

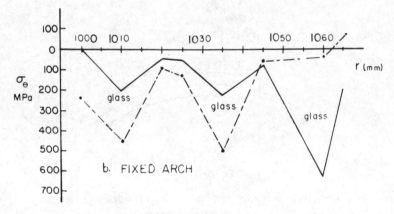

Figure 11. Stress distributions at the centre lines and at the fixed end of the composite arches of figure 10.

9.321° are modeled by 54 annular layered hybrid finite elements
(Fig 4). The elements representing the glass bands are 0.5 mm wide.
The circumferential bands have polyester cover thicknesses t = 1 mm
and a glass thickness c = 8 mm. The partial glass bands have a
thickness, c = 10 mm and t = 0. The elements representing the
matrix are 10 mm thick, 0.5 mm wide, at the inner and outer radii
of the ring, and are 1 mm wide everywhere else. All the annular
elements used are subtended by a small central angle of 1.552°.

Analysis of the composite rings under an internal pressure of
6.61 MPa shows a circumferential stress distribution pattern quite
different from that expected in homogenous rings without reinforcing
glass bands. The complete circumferential glass band, as shown in
Figs 6 and 9, carries a large proportion of the total loading,
about 34% under no failure conditions, with a stress of 400 MPa.
The band can carry the entire load in the post failure mode with a
stress of 1200 MPa. The 60 MPa stresses in the matrix adjacent to
the complete bands, are 60% of the values expected in a ring without
a complete reinforcing glass band. The presence of partial circumferential bands results in substantial increases in hoop stress in
the matrix, at the tips of these bands. An average stress increase
of 40% is observed in the ring of Fig 5 and an 80% increase is
observed in the ring of Fig 7. The presence of cavities results
in a redistribution of stresses in the matrix and glass bands.
Stress intensity factors are computed and plotted in Figs 6 and 8.
The intensity factors represent the ratios of the stresses in a
ring with cavities to the stresses in a ring without cavities.
The stress intensity factors are highest in the matrix regions that
are radially adjacent to a cavity and lowest in the matrix regions
that are circumferentially adjacent to a cavity. The highest
stress intensity values are 1.7 in the ring of Fig 6 and 1.9 in
the ring of Fig 8.

A nonlinear crack propagation analysis in a ring is shown in
Fig 9. Cracks AA'A" then BB'B" began in the neighbourhood of the
cavities and at the tips of the partial glass bands. The initiation and arrest cycles of the cracks were achieved in four
independent time steps. Crack CC' began at void C in a fifth time
step due to the very high strains (0.9%) in the adjacent circumferential glass band. Crack openings of 0.18 mm, 0.17 mm, and
0.13 mm were computed as the two sides of the ring were held
together by the circumferential glass band only. The brittle
nature of the polymer in the matrix[11] at cryogenic temperatures,
results in a catastrophic mode of failure.

The cryogenic stresses due to the differential thermal expansion of the glass bands and the matrix under cool down from 300 K
to 1.8 K are also analysed using an integrated contraction strain
of 135 x 10^{-5} for the polyester/chopped glass matrix and 22 x 10^{-5}
for the glass bands. Results show that the highest cryogenic

tensile stress of 9 MPa (8% of ultimate strength) in the matrix
occurs at the inner surface of the ring immediately adjacent to a
partial glass band and the highest compressive stress of -37.5 MPa
(2.5% of ultimate strength) occurs in the complete circumferential
glass band. These cryogenic stresses reduce the capacity of the
ring to carry internal pressures, and increase its capacity to
carry external compressive pressures.

CRYOGENIC COMPOSITE ARCHES IN A SUPERCONDUCTIVE STORAGE MAGNET

        Composite glass reinforced polyester structures, carrying the
rippled conductors of Energy Storage Magnets[1] are shown in Fig 1.
The arch supports are 1 m in radius and 0.065 m wide, 0.1 m thick
and have a central angle of 80°. The arches are reinforced with
three bands of alternate layers of circumferential glass 0.5 mm
thick and polyester/chopped fibres 2 mm thick. The arch construction is shown in Fig 10a. In a 0.1 m thick arch, the glass bands
are represented by 5 mm wide, symmetrically layered Hybrid
elements,[8] having total glass thicknesses of 0.044 m and total
matrix thicknesses of 0.055 mm. The arch is subjected to a
compressive radial loading of 12×10^5 n/m which is 1.5 times
higher than the pressure exerted by the conductor.

        Since the composite arch, in the structure of Fig 1, could be
either pin-ended or fixed-ended, both end conditions are investigated as shown in Fig 10b and c. Circumferential stress distributions along the crown width are shown for both types of arches,
and the stress distributions along the fixed end of arch 'c' are
shown in Fig 11. Results of the analysis indicate that the maximum
stresses in the three glass bands are comparable at the crowns of
arches 'b' and 'c' and at the fixed end of arch 'c' (625 MPa max.).
The maximum compressive stresses in the matrix of arch 'c' (250 MPa),
at the fixed end and at r = 1000 mm, is 38% higher than the maximum
compressive stresses at the crown (180 MPa) and r = 1065 mm. It
is concluded that an end condition for the arches closer in nature
to a pin would lead to a better design.

CONCLUSIONS

        Composite polyester/chopped fibre rings reinforced with
circumferential glass bands are analysed subject to internal pressure and cryogenic conditions. A crack propagation analysis leading
to failure of the ring is also performed. It is shown that the
values of stresses double in their intensities in regions adjacent
to cavities and at the tip of partial circumferential glass bands.
Designs for pinned and fixed-ended composite arches under compressive radial loadings are presented. Results indicate that a pin-
end joint would lower the maximum stresses in the arch matrix.

ACKNOWLEDGEMENTS

    The author is grateful for the generous support of the Depart-
ment of Energy and the Wisconsin Power Research Foundation, and
also to the support of the Department of Civil Engineering/Mat.
Sci. Eng. at the Johns Hopkins University. In addition, the author
sincerely appreciates the help and advice of R Boom, W Young,
R Green, and E Horowitz. The author is very thankful to K Han for
providing laboratory test data used in the analysis and to
J J Olazabal for his enthusiastic help and technical support during
the preparation of this work.

REFERENCES

1.   R W Boom et al, Wisconsin Superconductive Energy Storage
        Project, Vol. I (1974), Vol. II (1976), Annual Report
        (1977), University of Wisconsin, Madison, Wisconsin.
2.   E L Stone and W C Young, in: Advances in Cryogenic Engineering,
        Vol. 24, Plenum Press, New York (1978), p. 274.
3.   E L Stone et al, in: Nonmetallic Materials and Composites at
        Low Temperatures, Plenum Press, New York (1979), p. 283.
4.   S G Ladkany and E L Stone, in: Nonmetallic Materials and
        Composites at Low Temperatures, Plenum Press, New York
        (1979), p. 377.
5.   S G Ladkany, in: Advances in Cryogenic Engineering, Vol. 24,
        Plenum Press, New York (1978), p. 374.
6.   R W Boom et al, Wisconsin Superconductive Energy Storage
        Project, Vol. IV, University of Wisconsin, Madison,
        Wisconsin, to be published.
7.   K S Han and J Kontsky, "The Strain Energy Release Rate of
        Glass-Fiber-Reinforced Polyester Composites", presented at
        the ICMC Conference on Nonmetallic Materials at Low
        Temperatures, August 4-5 1980, CERN, Geneva.
8.   S G Ladkany, Five Hybrid Elements for the Analysis of Thick,
        Thin or Symmetric Layered Plates and Shells, Ph.D. Disser-
        tation AWB.L1572.5303, University of Wisconson, Madison,
        Wisconsin (1975).
9.   T H H Pian and Pin Tong, "Basis of Finite Elements Methods
        for Solid Continua", International Journal for Numerical
        Methods in Engineering, Vol. I, 1969.
10. P Pederson and M M Megahed, Computers and Structures, Vol. 5,
        Pergamon Press (1975), p. 241.
11. G Hartwig, in: Nonmetallic Materials and Composites at Low
        Temperatures, Plenum Press, New York (1979), p. 33.

THE STRAIN ENERGY RELEASE RATE OF A GLASS FIBRE

REINFORCED POLYESTER COMPOSITE

Kyung S Han and James Koutsky

University of Wisconsin
Madison
Wisconsin, USA

INTRODUCTION

A superconductive energy storage magnet requires a large
quantity of low-thermal-conductive, high-load-carrying supporting
structures to carry the radial and axial compressive loads from
the 1.8 K magnet coils to room temperature bedrock.[1-3] Engineering
data is needed on low cost commercial composites for struts to be
used in the superconductive energy storage magnet designs at the
University of Wisconsin-Madison. The main properties are compressive strengths, delamination strengths, shear properties and
compressive fatigue properties at cryogenic temperatures and room
temperatures.[4-7]

We are reporting here on the fracture energy of glass fibre
polyester composite. Usually the fracture energy for an adhesive
bond is measured or calculated for a rectangular cantilever beam
specimen. However, the fracture energy obtained varies with the
crack length which must be measured accurately. It is difficult
to measure crack length at room temperature and almost impossible
at cryogenic temperatures. A better sample is the width tapered
double cantilever beam (WTDCB) specimen. With this specimen there
is no need to measure the crack length either at cryogenic or room
temperatures. A width tapered double cantilever beam (WTDCB)
specimen was used for the fracture energy measurement of wood-
adhesive bonds[8] and metal-adhesive bonds.[9] Bascom et al[10] studied
the fracture behaviour of composites.

We review the width tapered double cantilever beam and its
application for the measurement of fracture energy at 300 K and
77 K. It is believed that this present work is the first to

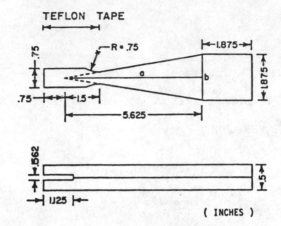

Figure 1.   Width Tapered Double Cantilever Beam (WTDCB)
            Specimen. Two 1/4" Samples Bonded Together.

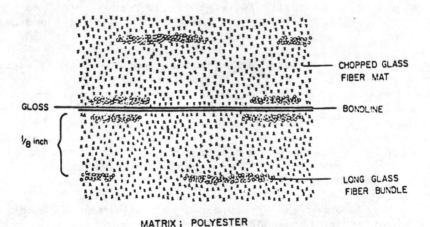

Figure 2A. Schematic picture of a cross section of a
           bonded Extren specimen.

utilise a width tapered double cantilever specimen at 300 K and 77 K to measure fracture energy.

THEORETICAL BACKGROUND

The strain energy release rate or fracture energy is defined as the rate of change of the stored strain energy of the beam with a change in the fractured bond area, ie,

$$G_{IC} = - \frac{dU}{dA} = \frac{1}{2} P_c^2 \left(\frac{dC}{dA}\right) \tag{1}$$

where G_{IC} is fracture energy, A is fracture area, U is potential energy, C is compliance and P_c is the critical load area at which the crack begins to propagate. The change in the fracture area is

$$dA = b(a) \, da \tag{2}$$

where b(a) is a width of specimen at the crack length a. After replacing dA in eq. (1)

$$G_{IC} = \frac{1}{2} \frac{P_c^2}{b(a)} \frac{dC}{da} \tag{3}$$

is used to determine G_{IC} from compliance measurements.

The basic differential equation for the deflection curve of a beam in pure bending, no shear deformation, is

$$\frac{d^2\varepsilon}{dX^2} = - \frac{M}{EI} \tag{4}$$

where ε is the deflection of the beam at a distance X from the end of the beam, M is the bending moment at X and equal to the load P times the moment arm X, I is the moment of inertia of the cross section and equal to $bH^3/12$, E is the modulus of elasticity of the beam material, and H is a thickness of the beam. Substituting into the differential equation for M and I gives

$$\frac{d^2\varepsilon}{dX^2} = - \frac{12PX}{EbH^3} \, . \tag{5}$$

Integrating the above equation twice with the boundary condition $\frac{d\varepsilon}{dX} = \varepsilon = 0$ at X = a at the fixed end of the beam gives

$$\varepsilon(X) = - \frac{2PX^3}{EbH^3} - \frac{6Pa^2X}{EbH^3} - \frac{4Pa^3}{EbH^3} \, . \tag{6}$$

The deflection ν at the loading end of a double cantilever beam is twice the deflection at $X = 0$ of a single cantilever beam. Thus

$$\nu = 2|\varepsilon(0)| = \frac{8Pa^3}{EbH^3} \quad . \tag{7}$$

The compliance of the specimen is

$$C = \frac{\nu}{P} = \frac{8a^3}{EbH^3} \quad , \quad \text{and} \tag{8}$$

$$G_{IC} = \frac{P_c^2}{2b(a)} \frac{dC}{da} = \frac{12P_c^2 a^2}{EH^3 b^2} \quad . \tag{9}$$

If taper $K = a/b$ is constant in eq. (9) then the strain energy release rate for a double cantilever beam (WTDCB) is independent of crack size.

For a WTDCB we get

$$G_{IC} = \frac{12P_c^2 k^2}{EH^3} \quad \text{and} \quad C = \frac{12\,Ka^2}{EH^3} \quad . \tag{10}$$

The only variable requiring measurement during testing is the critical load at the initiation of crack propagation.

These equations are derived assuming linear elastic behaviour without shear deformation and linear crack propagation. For such assumptions a plot of compliance versus crack length squared, eq. (10), should give a straight line with a slope of

$$m = \frac{12K}{EH^3} \quad . \tag{11}$$

EXPERIMENTS

Bond strength and fracture energy measurements of Extren,* a pultruded fibreglass polyester composite, are reported. Table I lists the mechanical properties of Extren. The Extren sample sheets are polished with an emery paper to remove the surface gloss and cleaned with acetone to remove sanding dust before bonding together with an adhesive. The adhesive is 10 parts Shell Epon 826 epoxy resin and one part curing agent diethylene tetramine which

---

*Manufacturer: Morrison Molded Fiberglass Co.

Table 1. Properties of Extren at 300 and 77 K

PROPERTY	300 K		77 K	
	LONGITUDINAL	TRANSVERSE	LONGITUDINAL	TRANSVERSE
TENSILE STRENGTH	200,000 psi	10,000 psi		
TENSILE MODULUS	2.0×10^6 psi	1.5×10^6 psi	2.4×10^6 psi*	1.8×10^6 psi*
COMPRESSIVE STRENGTH	35,000 psi*	29,000 psi*	52,000 psi*	38,000 psi*
FLEXURAL STRENGTH	25,000 psi*	18,000 psi*		
BOND STRENGTH	2380 psi*	2750 psi*		
IMPACT STRENGTH	20 ft-lb/in	6 ft-lb/in		
SHEAR STRENGTH	6,000 psi			
SPECIFIC GRAVITY	1.74 gm/cc^3			

* U.W. EXPERIMENTS

EXTREN MANUFACTURED BY MORRISON MOLDED FIBER GLASS COMPANY.
UNIDIRECTIONAL + RANDOM (44w/o) GLASS FIBRE REINFORCED POLYESTER
COMPOSITE.

Table 2. Fracture Energy of Extren

	300 K	77 K
Longitudinal	774 J/m^2	2200 J/m^2
Transverse	1084	3257

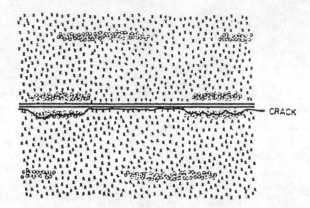

Figure 2B. A crack propagates through the glass fibre
bundle and mat portion

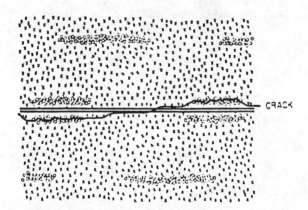

Figure 2C. A crack sometimes jumps over the bond line.

is mixed and poured onto both surfaces to be bonded. A precrack
at the loading end of the specimen results from a Teflon tape which
prevents bonding. The curing sequence is: 500 psi for 24 hours at
room temperature and followed by 6 hours at 500 psi at 106°C. The
plates are 3 mm, 9 mm and 12 mm thick. WTDCB specimens with a taper
of 1, 2, 3 or 4 are machined from the bonded plates. The compression and tension apparatus was described previously.[4,5] The load
cell is an Interface 1330-AF 100 K. A Linear Variable Differential
Transformer monitors the crack opening displacement of the beam.
Load (P) vs. displacement (ν) is recorded for a loading rate of
1.25 mm per minute. Identical tests are made at 77 K.

RESULTS AND DISCUSSION

Fracture Behaviour

     Cracks should propagate uniformly. Actually there are some
perturbations. First, the crack propagates by localised pop-in
at the initial stage which gives a low P_c. Second, because of
beam bending a thin specimen gives a curved slope in a load-
deflection (p,ν) curve which yields too high a bending energy.
Third, there are probably some plastic deformations at the notch
root.

     A bonded plate is required for the energy storage struts. If
crack propagation takes place along the bondline then the bond
material is weaker than the bulk material and the fracture energy
of the bond line is measured. If the crack propagates along the
bulk material, then the bond is stronger than the bulk material
and the bulk material fracture energy is measured.

     We find that cracks propagate through the bulk material just
below the bond line where the composite is the weakest. Figure 2
is a schematic picture of crack propagation through the glass
fibre bundle and mat portion (Fig 2-b) and sometimes the crack
jumps over the bond line (Fig 2-c). Cracks propagate easily
transverse to glass fibre bundles. Although crack propagation
modes are predictable it is very difficult to analyse the fracture
behaviour mathematically in the nonuniform structure.

     There are two kinds of crack propagation, stable and unstable,
which are sketched in Fig 3. Stable propagation has a constant
P_c with no crack arrest, Fig 3-a. The crack propagates conti-
nuously because there is no variation in structure. Unstable
propagation has crack initiation and arrest stages. Once the
crack propagates, the load drops and the crack arrests until the
next crack initiates. During the crack arrest period, there is
an increase in load, which is shown in Fig 3-b. The transverse
crack propagation follows this mode because it cuts across mat
and bundle portions and arrests at the tougher mat portion. The
mat is tougher than the bundle in the transverse direction.

Compliance Study

     Compliance specimens of tapers K = 1, 2, 3 and 4 and thicknesses
H = 6 mm, 9 mm, 12 mm and 18 mm are used. In one case, cracks are
inserted by sawing along the bondline with a thin 2.5 mm jigsaw
blade. Each specimen is first loaded and unloaded without propagating the crack to obtain a load-displacement curve at the precut
crack length. The crack was then advanced 18 mm using a jigsaw
and the procedure was repeated until the compliance versus crack
length data was obtained over a major portion of the tapered region
of each specimen. In the second case, crack propagation is initiated
by loading. Once the crack propagated the specimen was unloaded,
the crack front is marked on both sides of the specimen and the
procedure is repeated. The crack length is taken as the average of
both sides. Most of the compliance - crack length data is from
marking the cracks as they were arrested. The accuracy of the
crack length measurements is $\pm$ 0.75 mm. Compliance versus crack
length squared is plotted in Figs 4 and 5, with C calculated by
eq. (10) with E = 2 x 10^6 psi. Theory and experiment agree taper
k = 3 and H > 12 mm. As the beam thickness decreases, the experimental compliance moves farther from the theoretical values. This
compliance dependence on the specimen thickness might be explained
by the viscoelastic behaviour of the composite. During loading
there is a sudden decrease in load as a function of crack length,
Fig 6. The amount of hysteresis during the loading and unloading
cycle is also related to the crack length and dimension. Hysteresis
is thought to be a function of the area of the beam, ie crack
length squared and taper, Fig 7. However, if the area of the beam
is taken into account and calibrated the hysteresis data, the
corrected data fall on one straight line. This indicates that the
magnitude of the hysteresis depends on the area of the beam and
the crack length squared in a linear relation. This thickness
dependence on hysteresis is seen in Fig 8. The thinner beams have
larger hysteresis which decreases for thicker beams and finally
constant at thickness 12 mm.

     The compliance dependence on specimen geometry is a result of
viscoelastic behaviour of the beam. The optimum taper is 3 and the
minimum thickness is 12 mm for fracture energy tests of Extren.

Fracture Energy

     For an ideal case, the constant critical load for crack
propagation is in Fig 3. In our experiment, the crack propagates
at a lower load than P_c at the early stage of crack propagation.
G_{IC} could be obtained only beyond a certain critical crack length
which is a function of beam geometry. Thus G_{IC} gradually increases
as the crack length increases and finally reaches a plateau value
after a critical crack length as shown in Fig 13. The G_{IC} value
obtained in this study is the plateau value at a certain taper and

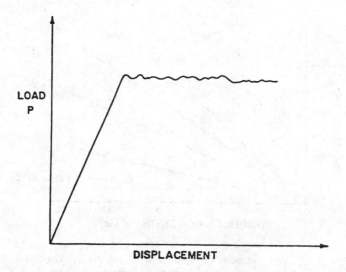

Figure 3A. Ideal Load Deflection Curves of a Width
Tapered Double Cantilever Beam Specimen.
Stable Crack Propagation.

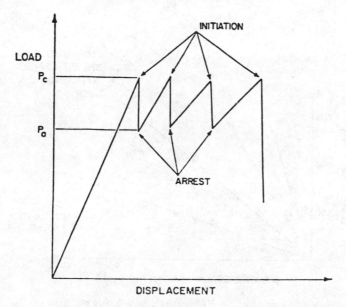

Figure 3B. Ideal Load-Deflection Curves of a Width
Tapered Double Cantilever Beam Specimen.
Unstable Crack Propagation.

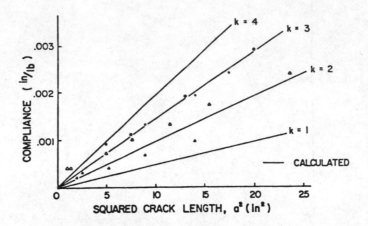

Figure 4. Compliance vs. (Crack Length)^2. Theoretical Line and Experimental Points Agree for K = 3.

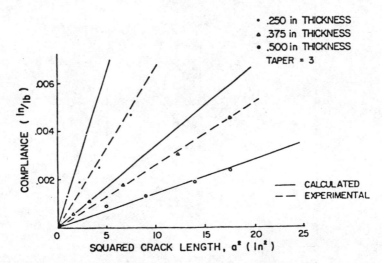

Figure 5. Compliance Versus Crack Length Squared. Theory and Experimental Agree for 0.5 inch Sample.

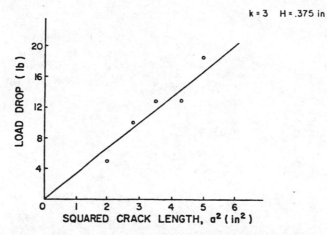

Figure 6.   Load Drop Versus Crack Length Squared. The
Amount of Load Drop During the Unloading
Procedure is a Linear Function of Crack
Length Squared.

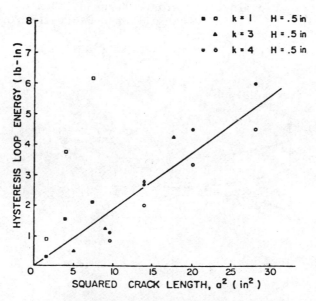

Figure 7.   Hysteresis Loop Energy Versus Squared Crack
Length Tapers. Open Marks are Experimental.
Closed Ones Calibrated. The Calibrated Data
Fall on a Straight Line.

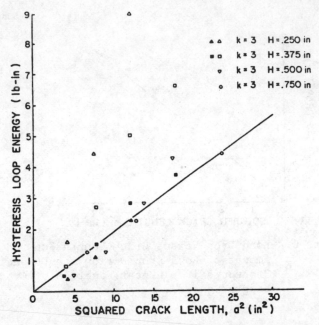

Figure 8. Hysteresis Loop Energy Versus Squared Crack
Length at Different Thickness. Open Marks are
Experimental, Closed Marks Calibrated. The
Calibrated Data Fall on One Straight Line.

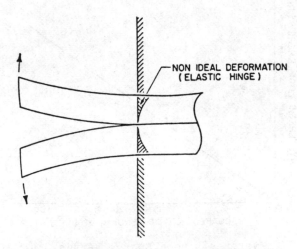

Figure 9. A Schematic Diagram of Nonideal Deformation
(Elastic Fringe) At the Crack Tip.

thickness.   One of the reasons for this nonideal behaviour is a
result of elastic fringe.   The cross section just above the crack
tip ideally undergoes no deformation of any kind, but the cross
section does deform as illustrated in Fig 9.   This nonideal elastic
fringe gives a large contribution at the early stage of crack
propagation.   As the crack propagates further, total volume of the
cantilever beam increases which compensates for the elastic fringe.
Therefore, the cross section just above the crack tip becomes less
deformed as the crack propagates.   These effects cause a low G_{IC}
at the early stage of crack propagation and a high G_{IC} at the later
stage of crack propagation.

   Figure 10 is a plot of the strain energy release rate versus
beam taper.   Although there is a large scatter between these values,
the G_{IC} value decreases as the beam becomes wider due to viscoelastic
behaviour.   The taper effect is more pronounced as the beam becomes
wider.   However, based on the compliance study and on the fact that
the crack front becomes more irregular as the taper becomes wider,
a taper of k=3 is the optimum value found for the fracture test of
the material used.

   Figure 11 shows the dependence of the strain energy release
rate on the beam thickness.   Ideally G_{IC} should be independent of
the beam geometry.   However, here it is seen that G_{IC} is large
below 12 mm thickness and becomes constant above 12 mm. G_{IC} obtained
from a thin plate is the sum of the fracture energy and the bending
energy.   As the beam becomes thicker, the bending energy becomes
smaller, and finally we obtain a reproducible fracture energy only
above 12 mm thickness.

Cryogenic Fracture Behaviour

   The same specimen geometry is used for tests in liquid nitrogen, a thickness of 12 mm and a taper of 3.   The relation
between the compliance and the squared crack length follows the
theory, ie, the compliance of the beam follows a straight line
relationship at cryogenic temperature as at room temperature,
Fig 12.   The longitudinal and transverse moduli obtained from the
slope of Fig 12 are 1.8×10^6 psi and 2.4×10^6 psi.   The moduli
increase at 77 K which is about 20% more than room temperature
moduli (1.5×10^6 psi and 2.0×10^6 psi).   The moduli results will
be used for calculating the elastic strain energy release rate at
77 K.

   Typical load-deflection curves at 77 K and 300 K are shown
in Fig 13.   The general load-deflection curves at 77 K behave in
a similar fashion as at 300 K.   The fracture energies at 77 K and
300 K are tabulated in Table 2.

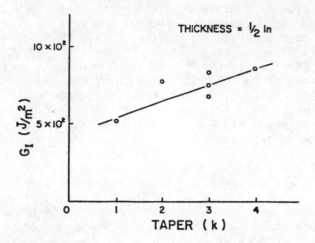

Figure 10. Fracture Energy Versus Beam Taper.

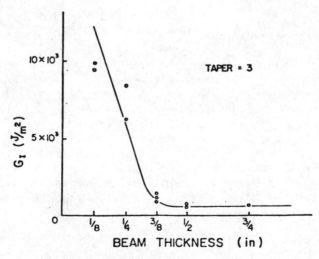

Figure 11. Fracture Energy Versus Beam Thickness.
Correct Value Above ½ Inch.

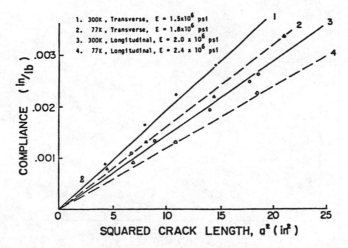

Figure 12. Compliance Versus Squared Length at 77
and 300 K.

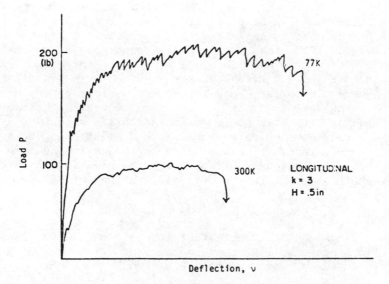

Figure 13A. Typical Load-Deflection Curves in
Transverse Direction.

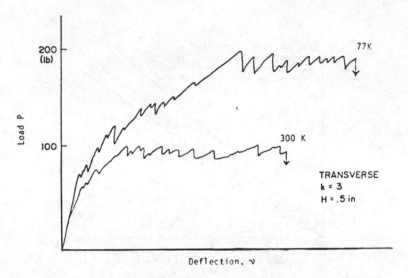

Figure 13B. Typical Load-Deflection Curves in
Transverse Directions.

        Generally metals undergo brittle transformation as the tempera-
ture decreases and toughness decreases although the strength
increases because of the ductility. However, Extren has about 3
times higher fracture energy at 77 K than at 300 K. The polyester
matrix material and the bonding between matrix and fibre affect
fracture energy. It is suggested that the matrix holds the glass
fibre better at 77 K because of the different thermal expansion
coefficients of matrix and fibre. The fracture surface at 77 K is
rougher than at 300 K. A rougher surface usually means a higher
fracture energy. However, scanning electron microscope pictures
show almost the same fracture surface in microscale. Unlike metals,
polymer changes its chain conformation and basic structure as the
temperature goes down which sometimes gives higher toughness at
cryogenic temperatures. The viscoelastic relaxation behaviour
greatly changes as a result of the chain conformation and structure.
This gives a higher fracture energy at 77 K than at 300 K.

Conclusions

    1. A Width Tapered Double Cantilever Beam (WTDCB) specimen
can be used successfully to measure the fracture energy of glass
fibre reinforced polyester composite at 300 K and 77 K.

    2. Crack propagation is dependent on the glass fibre direc-
tion; stable propagation along the glass fibre direction and unstable
propagation transverse to the glass fibre direction.

3. The compliance study shows that the optimum taper is 3 and the minimum thickness is one half inch.

4. Load drop and hysteresis studies show that the specimen behaves viscoelastically and hysteresis is a function of taper, thickness and crack length.

5. The fracture energy transverse to the fibre direction (1084 J/m^2) is slightly higher than along the fibre direction (774 J/m^2).

6. Fracture energy at 77 K (2200 J/m^2) is higher than at 300 K (774 J/m^2), suggesting that glass fibre reinforced polyester composite can be used successfully at a cryogenic temperature.

ACKNOWLEDGEMENTS

This work is supported by the US Department of Energy and Wisconsin Electric Utility Research Foundation.

REFERENCES

1.   R W Boom et al, Wisconsin Superconductive Energy Storage
        Project Report, Vol. I (1974), University of Wisconsin,
        Madison, Wisconsin.
2.   R W Boom et al, Wisconsin Superconductive Energy Storage
        Project Report, Vol. II (1976), University of Wisconsin,
        Madison, Wisconsin.
3.   R W Boom et al, Wisconsin Superconductive Energy Storage
        Project, Annual Report (1977), University of Wisconsin,
        Madison, Wisconsin.
4.   E L Stone and W C Young, Proceedings 7th Symposium on
        Engineering Problems of Fusion Research II, Knoxville,
        Tennessee, IEEE Publ. No 77CH1267-4-NPS (1977).
5.   E L Stone and W C Young, Advances in Cryogenic Engineering,
        Vol. 24, Plenum Press, New York (1978), p. 279.
6.   E L Stone et al, Proceedings of the ICMC Symposium on Nonmetal-
        lic Materials and Composites at Low Temperatures, Munich,
        West Germany, (1978), p. 283.
7.   R W Boom et al, Wisconsin Superconductive Energy Storage
        Project Report, Vol. IV (in press), University of Wisconsin,
        Madison, Wisconsin.
8.   M C Luce, M S Thesis (1980), University of Wisconsin, Madison,
        Wisconsin.
9.   S Mostovy, Communication from the 10th Interim Report,
        LR217614-10.
10. W D Bascom et al, Composites, Vol. II (1980), p. 9.

RECENT EUROPEAN WORK ON THE NONDESTRUCTIVE TESTING OF

COMPOSITE MATERIALS

W N Reynolds

Nondestructive Testing Centre
A.E.R.E., Harwell, England

INTRODUCTION

Fibre-reinforced composite materials, notably GRP and CFRP are
now finding increasing use in structural applications including
shipping, aircraft and aerospace as well as in large cryogenic
items. This development has revived interest in the special
problems associated with the testing and inspection of such materials
and it is therefore an appropriate time to survey existing methods
in relation both to their scientific background and their practical
shortcomings. Much work on the NDT of composites has, of course,
taken place in the USA but this paper aims to discuss only the main
European contributions.

Structural metals are frequently inspected for cracks, pores
and inclusions by well established NDT techniques but these methods
require modification in order to deal with composites which are
deliberately chosen for their comparative lightness and stiffness.
They also raise their own special problems and it has been found
profitable to develop tests which have no particular application to
metals.

Fibre and resin quality are regulated by sampling and process
control, but the use of composites raises problems such as the
determination of:

(a) State of cure of resin matrix

(b) Porosity of resin matrix

(c) Fibre volume fraction

231

(d)   Orientation and lay-up of fibres and plies

(e)   Fibre-matrix interface condition

and the detection of:

(f)   Delamination and translaminar cracks running parallel
      to fibres

(g)   Foreign inclusions

(h)   Lack of bonding of ply to adjacent ply or other structure

The significance of most of these factors is clear, but with
regard to (b) a recent survey has shown[1] that each 1% of voidage
in a composite material produces a decrease in shear strength of
7% at least up to 4% of voids. Typical users therefore commonly
specify an upper limit of 0.5 - 1% voids by volume. Local
delaminations between plies are thought to become significant[2] as
they exceed a length of the order of 10mm.

Current Techniques

Although many techniques of different types have been proposed
for the post-fabrication assessment of composites, the industry
relies at present almost entirely on a group of well established
traditional methods. Thus visual inspection for fibre alignment
in CFRP or porosity in GRP is taken very seriously and coin tapping
by experienced inspectors is widely employed as a general check
for defective areas. The general test for high performance structures is the ultrasonic C-scan, backed up where necessary by the
Fokker bond tester adapted to a slightly different role, and also
by high resolution low voltage X-radiography.

Ultrasonic C-scan

This is best carried out by a total immersion technique as
illustrated in Figure 1, showing equipment developed along lines
described by Jones and Stone[3] for the inspection of laminate panels.
The panel to be inspected is supported above a flat reflecting surface of metal or glass and an ultrasonic transducer is used in a
pulse-echo mode to record the amplitude of the reflected signal.
The transducer may be of the focused variety or, as in the diagram,
the signal/noise performance may be improved by the use of
collimating stops. The frequency employed is usually in the range
5 - 10MHz, depending on the thickness of the laminate to be
inspected, and the carriage is traversed on an x-y scanning plane
with suitable instrumentation to produce a permanent record on
Mufax paper. This recording is usually 'quantised' so that areas
with high attenuation are clearly outlined.

The defects detected by this method are delaminated areas,

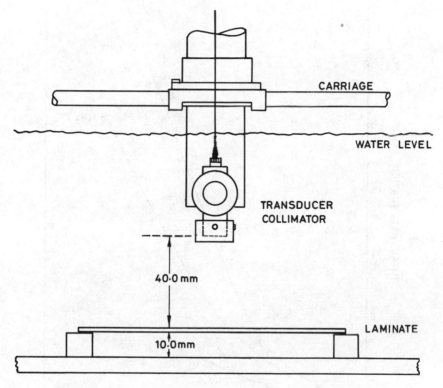

Fig. 1.   Collimated beam C-scan equipment

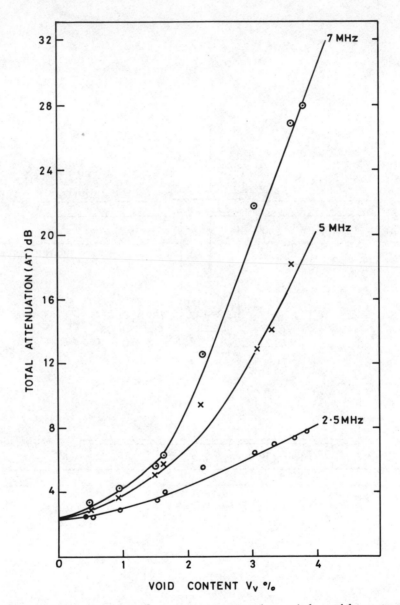

Fig. 2. Variation of total attenuation with void content

which have a very high attenuation and areas of porous matrix which
have comparatively high attenuation. For a relative porosity p of
pores of mean radius a the attenuation α varies as $a^3 p/\lambda^4$
if the frequency used is such that $k\,a < 1$. Calibration experiments are performed for each resin-fibre system and measurements made of reflective losses at each interface so that the
porosity and its variation are accurately known over the entire
scanned areas. This porosity is also carefully standardised
against measurements of interlaminar shear strength. Some results
given by Jones and Stone are illustrated in Figures 2 and 3. It is
possible, on the basis of data of this kind, for an agreed minimum
interlaminar shear strength to be specified in terms of maximum
pore content and, to a lesser extent, on maximum permitted delamination size. In practice it is not always convenient to immerse
the composite part in a water tank. Much working space can be
saved by the use of pairs of water coupled probes mounted on
callipers to make measurements in transmission. In satellite parts,
where water contamination of the surface must be avoided, Crocker
and Bowyer[4] have advocated the use of oil filled rubber tyred roller-
probes with suitable spring mountings.

Spot checks for delaminated areas may also be carried out by
means of pulse-echo equipment using a hand-held probe of the ultrasonic thickness gauge type. A local couplant is required and
delaminations are indicated by anomalously low readings.

Low Voltage X-radiography

X-radiography of fibre reinforced plastics can be very
informative if carried out under suitable conditions. Low voltage
equipment (7 - 35kV) with a fine focus source (15µm) and a Beryllium
window is the most successful, with single sided photographic
emulsion recording. The exposures may be made under either contact
or projection conditions depending on the area to be covered,
resolution required, cost etc. This technique is particularly
useful for detecting inclusions and fibre misorientation and for
that purpose a glass tracer fibre is sometimes included in carbon
fibre tows. Somewhat higher voltages (30 - 50kV) are employed for
the examination of honeycomb composite panels, to detect distorted
or unbonded areas.

Figure 5 is a survey of X-ray data obtained by Tober and
Schnell[17] using the contact method. Sensitivity can be improved by
projection method.

Fokker Bond Tester

This is a commercial ultrasonic resonance device which is widely
used in the aircraft industry to detect certain adhesive bond
defects, such as porosity, incorrect cure of the adhesive, or in-

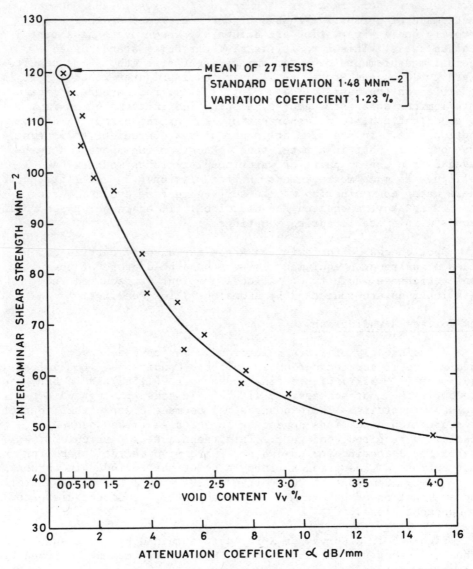

Fig. 3. Relationship of shear strength to attenuation coefficient and void content

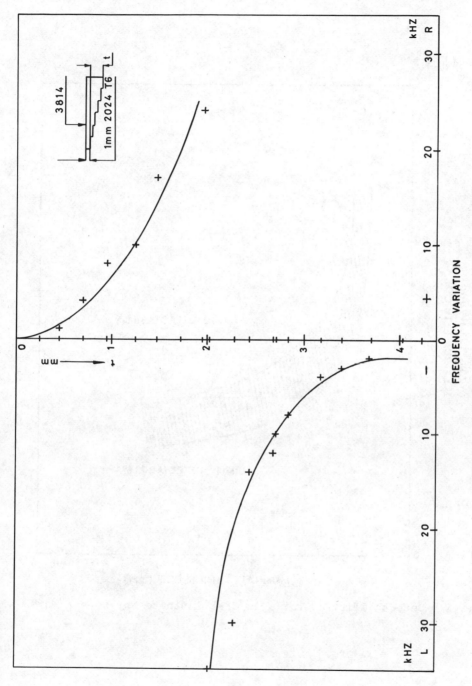

Fig. 4. Fokker bond tester readings with thickness of CFRP.

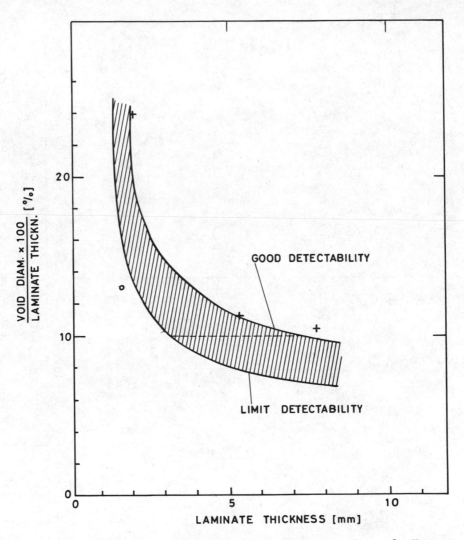

Fig. 5.   Detectability of voids in CFRP laminates by soft X-rays.

correct glue-line thickness. It cannot guarantee bond strength but
its use in composite panel testing is complementary to conventional
ultrasonics in that it uses shear modulus and damping coefficients
as indicators. Recent theoretical work[5] has shown how this
information can be combined with compressive wave data to give
precise data on composite quality. Figure 4 illustrates the change
in bond-tester reading with laminate thickness up to 4mm, or
apparent thickness in the presence of voids or delaminations.

The Fokker bond tester requires the use of a fairly stiff
couplant and cannot be used in a scanning mode but only as a point
by point tester. Other instruments operating on somewhat similar
principles are the Acoustic Flaw Detector, which is less sensitive
but does not require a couplant and the Shurtronic Sonic Resonator.

Outstanding Problems

The techniques of the previous section have been found generally
satisfactory for the examination of materials and structures before
use, but there are several problems remaining to which only partial
solutions have yet been developed. These are notably:

Examination of complex shapes with double curvature,
limited access or convoluted surfaces. In this case it has
been proposed to use measurements of ultrasonic wave velocity
either between separate transmitter and receiver placed at
fixed points or in a pulse-echo thickness gauge arrangement.
Theoretical and experimental studies[5] have shown how the
measured velocities can be related to fibre content and matrix
porosity. For a complete analysis it may be necessary to
combine shear wave velocities with compressive values or to use
density measurements from gamma-ray attenuation constants but
for inspection purposes it is normally only necessary to show
that the measured values agree within given limits with those
calculated or measured on a satisfactory structure. Gamma-ray
work at AERE[6] has shown that fibre contents may be rapidly
assessed to ±1% by the use of a small[238] Pu source of 17keV
photons. Other work not published in detail, has confirmed
that GRP and also CFRP laminate thicknesses can be measured
rapidly by the use of a commercial 200KHz eddy current device
in conjunction with a conducting backing foil. The conductivity
of the CFRP is too low to affect readings for laminates up to
20mm in thickness at least.

Effects of Fatigue

NDT procedures appropriate for the detection of mechanical
fatigue have been described in two recent papers[7,8]. These
suggest that significant factors additional to those already
discussed include the initiation of small shear cracks and

delaminations, the water content of the resin, and disbonding of fibres from the matrix.

The length of a surface opening crack can be determined by the use of zyglo penetrants with ultra-violet fluorescence, and the depth can also be measured by radiography using a suitable radio-opaque penetrant. Materials used successfully include liquid sulphur, carbon tetrachloride, trichlorethylene, di-iodo-methane and tetra-bromo-ethane, but they have not been deployed in an industrial scale partly because of possible health hazards.

Fibre matrix disbonding which is a symptom of both early fatigue damage and environmental degradation can at present only be measured by instrumented proof tests using changes in complex compliance or acoustic emission. So far these have been demonstrated only on simple laboratory specimens[9,10]. In GRP, damage can also be localised by simultaneous observation of infra-red emission[11] or by detailed dynamic analysis[12].

Recent work by Evans[13] suggests that thermal cycling up to 20 times between ambient and liquid nitrogen temperature increases ultrasonic attenuation and decreases interlaminar shear strength in a somewhat similar way to stress cycling. Although there are many difficulties and complications associated with the observation of stress wave or acoustic emission, it is possible that this is a case where in-service monitoring by this technique would be appropriate. The Kaiser effect would perhaps ensure that no new noise was detected after the first one or two cycles unless mechanically significant damage was occuring. There is, however, a regrettable lack of discussion of temperature cycling effects in the literature.

Water Content and Environmental Degradation

Although these factors are known to affect composite performance including fatigue there has been no systematic European work published on their measurement or nondestructive evaluation.

Rapid Inspection of Complex Structures

The foregoing account shows clearly that there is a need for rapid inspection or scanning, preferably without physical contact of complex structures. Various schemes of acoustic emission, mechanical tapping response, infra-red transmission or scanning, optical laser holography or speckle photography have been proposed, but so far none has reached the stage of industrial exploitation; although schemes for the industrial evaluation of laser holography have been described by Querido[14]

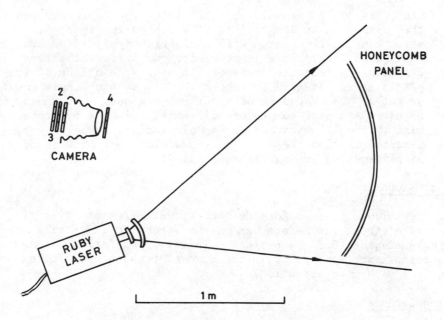

Fig. 6(a).   Ruby laser speckle phtograph with multiple emulsion
             camera to separate distortions

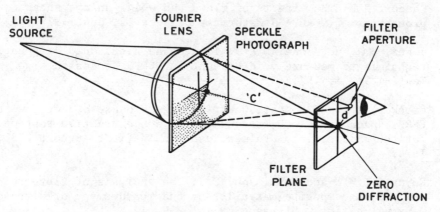

Fig. 6(b).   Fourier transform system for the analysis of defocus-
             sed speckle photographs

and Treca[15]. An example of a promising speckle technique is
that described by Gregory[16] , illustrated in Figure 6. The
structure is illuminated with a divergent ruby laser beam and
photographed by double exposure for two different stress
states using defocused cameras. By suitable choice of imaging
planes it is possible to record the transverse displacement
or out of plane bending of specified regions. The photographs
obtained are used to produce diffraction images by convergent
mercury arc illumination. The calculations required for
defect detection are at present unwieldy, but methods of
simplification are now being studied.

Conclusions

Techniques evolved for the nondestructive examination of
composite structures before use in the aerospace industries are
largely adequate for the purpose. The study of fatigue damage and
its assessment, especially in the case of temperature cycling,now
requires much more attention.

Acknowledgements

The literature survey on which this review is based was
supported by the European Research Office of the United States Army.

REFERENCES

1. Judd, N C W and Wright, W W: Voids and their effects on the
    mechanical properties of composites - an appraisal, SAMPE
    Journal 78, 10, 1978

2. Hancox, N L: The effects of flaws and voids on the shear
    properties of CFRP. J Materials Science 12, 884, 1977

3. Jones, B R and Stone, D E W: Towards an ultrasonic attenuation
    technique to measure void content in CFRP, Nondestructive
    Testing 9, 71, 1976

4. Crocker, R L and Bowyer, W H: The significance of defects in
    CFRP bonded honeycomb structures and nondestructive test methods
    for their detection. Fulmer Research Institute Report R736,
    1978.

5. Reynolds, W N and Wilkinson, S J: The analysis of fibre-reinfor-
    ced porous composite materials by the measurement of ultrasonic
    wave velocities. Ultrasonics 16, 159, 1978

6. Smith, A and Reynolds, W N: Use of a low energy gamma source to
    monitor CFRP. AERE Report - NDT/50, 1971

7. Sturgeon, J B:Fatigue mechanism, characterisation of defects
    and their detection in reinforced plastics materials. Brit
    J NDT 20, 303, 1978

8. Reynolds, W N: Fatigue evaluation of fibre-reinforced composite
    materials. AERE-R9439 to be published in Proc 16th Int Conf on
    aircraft fatigue, 1979

9. Adams, R D and Flitcroft, J E: Assessment of matrix and interface
    damage in high performance fibre reinforced composites. Proc
    8th World Conf on NDT, Cannes 1976, Paper 4B3

10. Adams, R D and Flitcroft, J E: The detection of matrix cracks
    in fibre reinforced plastics and their effect on strength.
    Proc ICE Conf on fibre reinforced materials, London 1977

11. Pye, C J and Adams, R D: Thermography as an NDT tool for
    composite materials. 9th World Conf on NDT, Melbourne 1979,
    paper 5B3

12. Cawley, P and Adams, R D: A Vibration technique for NDT of
    fibre composite structure. J Composite Materials 13, 161, 1979

13. Evans, D and Morgan J T: Physical properties of epoxide resin/
    glass fibre composites at low temperatures. Proceedings of
    ICMC, topical conference, Geneva, August 1980 (Plenum Press,
    New York)

14. Querido, R J:Holographic NDT of advanced composite materials
    in aerospace constructions. 8th World Conf on NDT, Cannes
    1976, paper 3A6

15. Treca, M: State of the art and evolution of a method for
    testing carbon composite aircraft structures. Proc 45th AGARD
    Structures and Materials Panel, Voss 1978, AGARD CP234, paper 18

16. Gregory, D A: Laser speckle photography and the sub-micron
    measurement of surface deformation on engineering structures.
    NDT International 12, 61, 1979.

17. Tober, G and Schnell, H: Detectability of flaws in boron and
    carbon composite parts. Proc 45th AGARD Structures and
    Materials Panel meeting. Voss 1977, AGARD CP234, paper 17

PHYSICAL PROPERTIES OF EPOXIDE RESIN/GLASS

FIBRE COMPOSITES AT LOW TEMPERATURES

D Evans and J T Morgan

Rutherford and Appleton Laboratories
Chilton
Didcot, Oxon OX11 0QX

INTRODUCTION

The versatility of epoxide resins may be regarded as both an
asset and a disadvantage in terms of low temperature applications.
An asset because the vast range of materials available, together
with all the formulating variables possible, enables most problems
to be overcome. A disadvantage because this bewildering choice
often leads to the selection of the wrong material with consequent
processing/manufacturing problems or mismatched physical properties
in the cured state.

Frequently the choice of resin system must be a compromise,
based on physical properties of the cured material and the processing characteristics at the manufacturing stage[1]. Here again, the
choice of resin system to suit the required manufacturing process
is a vital one and a knowledge of the likely physical properties
of the finished product cannot always be derived from conventional
testing and test methods. The very versatility that makes epoxide
resins so attractive again may be regarded as a disadvantage in
that the range of production techniques possible may give rise to
unexpected physical properties. It is essential that when considering the properties of composite materials, due regard is paid to
such variables as glass (filler) content, direction of glass fibres
relative to test direction and resin or filler type[2]. It is for
these reasons that this paper attempts to draw on our practical
experience of using these materials in low temperature environments[3]
and to relate this experience to the properties of other structural
materials and a range of physical test results, often achieved using
non-standard methods of test.

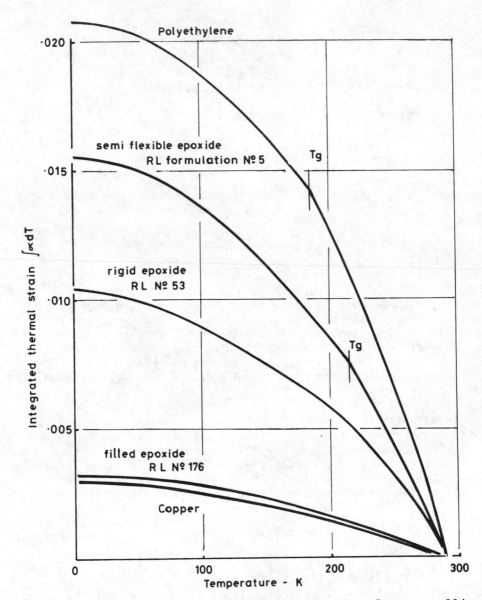

Fig. 1.   Thermal contration integrals for a range of non-metallic
          materials (shown relative to copper).

PHYSICAL PROPERTIES

Thermal Contraction

The range of thermal contractions[4] to be expected when considering plastics materials in relation to copper is shown in figure
1. It is important to note the differences even between epoxide
resins, there being no typical figure for an epoxide system.[5] Rather
there is a range of contractions which may be typified by a range
of epoxides. The soft, more flexible epoxides exhibiting an integrated contraction figure similar to that of polyethylene, while
the rigid epoxides have a much lower overall contraction with the
tough materials being somewhere between. In general, the lower the
brittle temperature the higher is the overall integrated contraction
at 4.2K. The reason for this is again apparent from figure 1, where
it may be clearly seen that at temperatures below the brittle
temperature (tg), most plastics materials behave in a similar manner,
with similar contraction coefficients. However, above the brittle
temperature the contraction coefficient is much increased and large
thermal strains are likely for relatively small temperature changes.

The incorporation of fillers into epoxide resins further modi-
fies the thermal properties but the processing characteristics, in
the liquid state, are also modified. Again, there is not a typical
figure that may be quoted to represent the effect of filler concentration[1] on thermal contraction; fillers affect the viscosity of an
epoxide resin system in accordance with the particle size and shape,
as shown in table 1. The smaller the particle size the greater the
filler surface area and therefore the greater effect of a given
loading on viscosity. The influence of particle shape on the
processing characteristics of liquid epoxide resins is also dramatic.
Note in table 1 how relatively low concentrations of chopped glass
fibres affect the viscosity and how the viscosity varies with the
fibre aspect ratio. Inevitably, the filler selected must represent
a compromise, based on the desired method of processing and the
physical properties of the end product. The machining characteristics of the cured material are yet another variable, though often
sacrificed in the effort to satisfy other parameters. (See table 1.)

The thermal properties of glass fabric/resin composites
require careful consideration prior to incorporation of such
materials in cryogenic equipment. When particulate fillers are
used in conjunction with epoxide resins a material results that
exhibits physical properties uniform in all directions. This is
not the case with glass fabric composites where the union of resin
with layers of woven fabric results in a highly anisotropic
material. Normal to the thickness of the laminate the thermal
behaviour of the composite will be strongly influenced by the
continuous resin phase and less affected by the discontinuous glass
phase. (See figure 2.) It is only at high glass contents, above

Table 1. Physical Characteristics of Various Filler Particles

FILLER	PARTICLE SHAPE	AV PARTICLE SIZE μm	S.G.	BULK DENSITY	MAX. CONC. FOR POURABLE MIX. VOL %	MACHINING CHARACTER- ISTICS
Zirconium Silicate	granular	50	4.6	1.65	59	Poor
Talc	acicular & granular	8	2.8	0.61	42	Fair
Aluminium Oxide	granular	<40	3.75	0.74	46	Poor
Woolastonite	acicular	3	2.9	0.73	46	Fair
Silica	granular	40	2.65	1.02	55	Poor
Glass	acicular	90 x 12	2.6	0.97	41	Poor
Glass	acicular	150 x 12	2.6	0.80	31	Poor

Table 2.

LAMINATE NUMBER	LAMINATE TYPE	FABRIC CONSTRUCTION	THERMAL CONTRACTION			
			AXIAL $\int \alpha dT$*		CIRCUMFERENTIAL*	
			77K	4.2K	77K	4.2K
1		plain weave (0.15 mm)	0.0035	0.0040	0.0035	0.0040
2		uni-directional	0.0050	0.0055	0.0016	0.0018
3		cross-plied, uni-directional	0.001	0.0012	0.0015	0.0018
4		woven roving	0.0030	0.0034	0.0030	0.0034

* refers to laminate direction relative to roll.

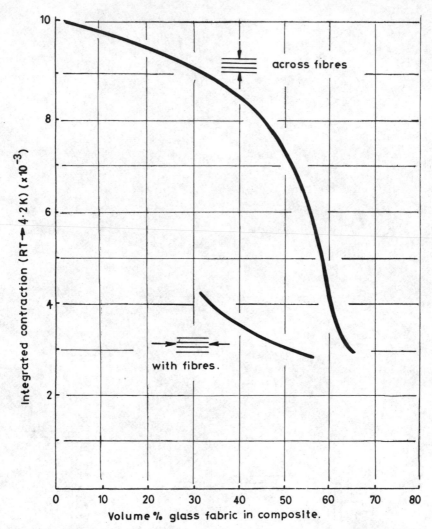

Fig. 2.   Thermal contraction for glass fabric/epoxide resin composite
          as a function of glass content.

the 40 volume % achieved in many good quality conventional composites, that the presence of the glass demonstrates a significant
effect. When considering thermal properties that the composite may
exhibit in the fibre direction, it is readily apparent from figure
2 that resin content has less influence on the overall contraction.
However, for the two principal directions representing the weave of
the fabric the glass content may vary, as for example in a warp
directional fabric, therefore differences in contraction may be
expected depending on the relative glass content in each direction.

In many instances the application of glass fabric/resin composites is in the form of rings or tubes and in these circumstances
the anisotropic nature of the properties is of paramount importance.
The thermal behaviour of these shapes is summarised in figure 3,
the measurements only being made at 77K because of the difficulty
in handling the relatively large structure involved. It was found
that there was a limiting thickness, relative to the diameter, that
could be successfully and reliably cooled to the temperature of
liquid nitrogen and that this thickness to diameter ratio was
significantly smaller than would be predicted from calculations
based on the known mechanical and thermal properties of the
material. Consideration of the thermal properties of rings in
structures where a resin/glass composite is used in conjunction with
metallic structures leads to additional problems; for example, when
rings of GRP are adhesively bonded to, say, stainless steel structures, the differential contraction integrals may work against
successful bonding, as shown diagrammatically below.

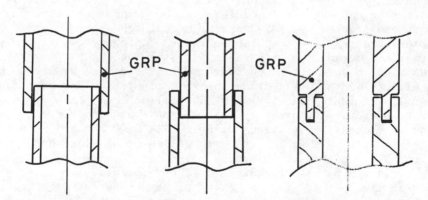

(a) bond may fail on   (b) bond may fail on   (c) sandwich construc-
    cooling because   cooling because   tion to minimise
    s/s contracts   o.d. of GRP con-   possibility of
    more than i.d.   tracts more than   failure.
    of GRP tube.   stainless steels.

Having experienced bond failures for the reasons outlined
(a and b), we have achieved successful marrying of the two materials
using the joint design shown at 'c'. The thin section of the GRP
imposes less stress on the bond line and the sandwich structure
also contributes to an overall lowering of thermal strain.

It might be assumed that regardless of the thickness:diameter
ratio of a GRP ring, the contraction through the thickness would
remain constant. This is in fact not the case (see figure 3)
because of the anisotropic effects previously mentioned. When the
ring is infinitely thin, ie the thickness to diameter ratio is low,
then the contraction through the thickness is much as would be
expected. The contraction through the thickness gradually decreases
with increasing thickness:diameter ratio and this must inevitably
mean that thermal strains are accumulating within the material.
Catastrophic failure by delamination occurs at a thickness to
diameter ratio of approximately 0.15, and at this point the measurements of variations in thickness induced by temperature changes
become unreliable.

Thermal Cycling

In cryogenic equipment there are instances where non-metallic
materials may be subjected to a large thermal gradient, with the
possibility of one end of a component working in liquid helium with
the other end at normal room temperature. In addition the operation
may be intermittent, involving a number of cycles between the upper
and lower temperatures. This mode of operation has not been investigated for resin systems formulated with particulate fillers but some
information has been obtained for various composites involving glass
fabric and epoxide resins. Four principal types of composite panel
were prepared, as shown in table 2, with the object of investigating
thermal and mechanical properties and studying changes in these
properties as a result of thermal cycling. To date we have compared
the effects of thermal cycling on two laminate systems by measuring
the inter-laminar shear strength (I.L.S.) at room temperature after
various numbers of thermal cycles. We have also compared this
measured mechanical behaviour with results from a non-destructive
method of testing based on the attenuation of an ultrasonic pulse.
Laminates 3 and 4 (table 2) were subjected to thermal cycling and
then evaluated by both techniques. The results are shown graphically in figure 4 and for the cross-plied uni-directional laminate
(number 3) some results are also presented in table 3.

It should be noted that after two cycles the inter-laminar
shear strength reaches a value that remains relatively constant.
This trend is not reflected by the N.D.T. results which appear to
show damage increasing with the number of thermal cycles.

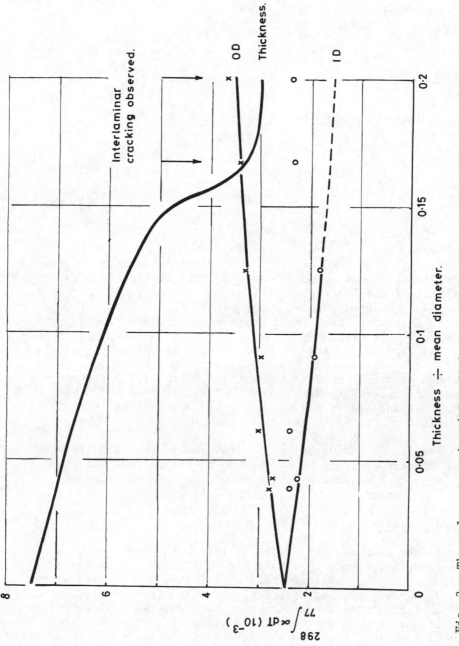

Fig. 3. Thermal contraction (RT → 77K) for glass fabric epoxide resin composite tubes.

Table 3.

No. of thermal cycles (RT>77K)	Attenuation (dB)	I.L.S. (MN/m^2)
0	0	51
1	1.5	33
2	2.1	21
3	2.8	24
4	5.3	18
10	7.0	22
20	9.0	19

Laminate 3 was selected for examination since each layer consisted of uni-directional fibres but the angle of the fibres differed in each subsequent layer by 22.5°. It was therefore considered that since each layer would have different thermal properties in the axial or circumferential direction of the laminate than adjacent layers, progressive or long term damage may result from thermal cycling. Damage from this source may be reduced in a laminate with uniform properties in each layer, and therefore results from laminate 3 were compared with laminate 4, where both mechanical and N.D.T. methods indicate a lower incidence of degradation.

For the N.D.T. method water was used as the coupling agent, but elaborate precautions were taken to ensure that moisture penetration into damaged areas of the specimen did not lead to confused results.

Fatigue Testing

It had been noticed, empirically, that glass fabric epoxide resin composites showed visible signs of damage after a relatively small number of mechanical fatigue cycles at low temperatures. This damage appeared to be restricted to surface crazing, and short term mechanical tests at a range of temperatures failed to show significant changes in strength or stiffness. It was also observed, in moulded G.R.P. components, that crazing appeared in areas where the material was subject to tensile stresses. Where compressive stresses of a similar magnitude were present crazing was not

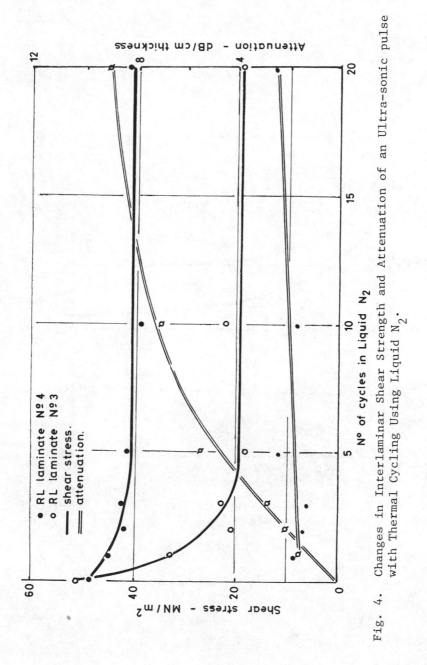

Fig. 4. Changes in Interlaminar Shear Strength and Attenuation of an Ultra-sonic pulse with Thermal Cycling Using Liquid N_2.

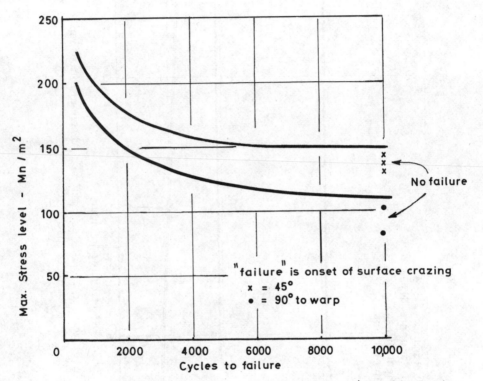

Fig. 5.   Flexural Fatigue Tests on Glass Fabric/Epoxide Resin
          Composites (Marglass Spec. 116T)

observed. While this crazing did not seem to have affected the
mechanical integrity of the composite it was regarded as the precursor of more serious damage which would possibly grow in use or
represent a possible source of trouble during thermal cycling.
Moisture ingress into such damaged areas could result in further
degradation on cooling.

Some data[6] already exists for epoxide resin composites at low
temperatures - but the criterion taken to represent the failure
point in conventional fatigue testing provides scant information
for the designer to whom crazing may be significant. A series of
laminates was manufactured and many specimens were prepared for
testing at a series of low stresses in a 3-point flexural bend
test at 77K. In this configuration both tensile and compressive
stresses are present and therefore the incidence of tensile or
compressive failure could be examined.

The results are shown graphically in figure 5, where it should
be noticed the ease with which surface crazing could be induced.
At stresses as low as 10% of the low temperature ultimate, fatigue
crazing could be induced in only a few thousand cycles. Indications
are that varying the resin system has little effect on this crazing,
but we have not yet investigated the influence of different types
of glass fabric. In all our investigations specimens 'failed' in
tension, presumably because of the unyielding nature of the resin
when stressed in this mode.

CONCLUSIONS

Confidence in the use of glass fabric reinforced epoxide resin
composites at low temperatures is slowly being established. Long
term 'use experience' together with information derived from tests
designed specifically to help understand the low temperature
behaviour of these materials has been used to complement purely
mechanical test results. The state of the art is now such that
predictions may be made, with some confidence, about the likely
performance of a range of composite materials in a cryogenic
environment. It is likely that quality control will be a significant factor in the long term use of non-metallic materials, because
of the need to employ only high quality materials for reliable low
temperature performance. Early indications are that further
changes in the resin phase will achieve only marginal improvements
in composite performance and that the nature of the reinforcement,
manufacturing technique and quality control will offer the most
rewarding paths to reliable performance.

The selection of materials, based on an in depth understanding of all factors involved in both manufacture and operation, will also form an important aspect of the future use of non-metallic materials.

REFERENCES

1. Evans, D, Langridge J U D, and Morgan, J T: Proc ICMC - Munich
    (1978).

2. Evans, D, and Morgan J T: 14th Int Conf on Reinforced Plastics,
    Paris (1979)

3. Evans, D, Langridge, J U D, and Morgan, J T: Proc ICMC, Madison
    U.S.A. (1979)

4. Van de Voorde, M: IEEE Trans on Nuclear Science, Vol 20 (1973)
    p693-7

5. Evans, D, Morgan, J T and Stapleton, G B: Rutherford Laboratory
    report RHEL/R251 (1972)

6. Kasen, M B: Cryogenics June 1975, p 327-349

CHARPY IMPACT TEST OF CLOTH REINFORCED

EPOXIDE RESIN AT LOW TEMPERATURE *

S Nishijima and T Okada

Department of Nuclear Engineering
Osaka University, Suita
Osaka 565, Japan

1. INTRODUCTION

Extremely large superconducting magnets have been recognised as essential in projects such as energy storage or fusion. The increase in size and pulsative operation of such magnets makes applied forces in components complex and varied. It is important, therefore, to elucidate the behaviour of each magnet component against the various forces at cryogenic temperature.

Organic materials, especially epoxide resin, have been employed as insulators and/or potting materials in the medium sized super-conducting magnet because they show easy fabrication, good insulation characteristics and good adhesion at cryogenic temperatures. Moreover, the fibre reinforced plastics with epoxy matrix are expected to have excellent mechanical properties not only as insulators but also as structural materials. In contrast with many other applications, the possible shortcomings of these materials have not yet been fully ascertained; in particular, these materials are sensitive to temperature and/or strain rate.[1-4]

The selection criteria for cryogenic materials vary with the application; among them dynamic toughness will be one of the most important. The dynamic toughness is defined as the ability for energy absorption or plastic deformation in the process to fracture under impact loading. To increase reliability and endurance of the pulsating machine it is indispensable for the structural materials to possess sufficient toughness.[5] Though we might point out some

* This work is supported in part by the Grant in Aid for Scientific Research No 504533, Ministry of Education in Japan.

other important properties, materials lacking toughness will not be
accepted as structural materials even if they satisfy other requirements. While tensile, compressive, flexural and hardness tests
offer useful information, they do not indicate the response of the
material to concentrated and impact loading.

In the present work the authors have chosen an epoxide and an
FRP with epoxide matrix as samples and attempt to establish a
method for obtaining dynamic toughness and to provide basic data
for design, using the Charpy impact test at cryogenic temperatures.
They also try to reveal the failure mechanism and the potential
problems associated with the use of these materials in practical
applications in superconducting magnets.

Uninstrumented Charpy test is not a fundamental test and
hence the values obtained cannot be substituted in fracture equations.
The values obtained by instrumentation also cannot be substituted
in fracture equations because FRP itself indicates nonlinear
fracture behaviour. The authors are now considering Charpy impact
values in this work as a precursor to the use of an instrumented
Charpy test.

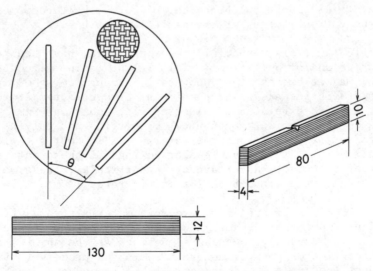

Figure 1. Orientation and specifications of specimens cut
from disc.

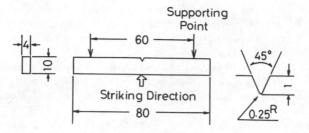

Figure 2.   Shape and dimensions of test specimen

2.   EXPERIMENTAL

2.1   Samples

The samples are epoxide resin, glass cloth reinforced epoxide resin and carbon cloth reinforced epoxide resin. The resin is a commercial epoxy resin (6861-2) composed of Bisphenol-A and is supplied by Ryoden-Kasei Co Ltd. This resin[6] is cured by triethanolamine-titanate-triphenylborate complex with a small amount of diluent (phenyl glycidyl ether). Cure temperature is 150°C for twelve hours. The glass transition temperature of this epoxide resin ranges from 50 to 60°C. Three kinds of sample are prepared. One is cured without reinforcement (named "Sample A"). The second is the glass cloth reinforced epoxide resin with 60 plys by hand-lay-up method (called "Sample B"). The glass cloth used in Sample B is plain E-glass which shows 350 kg/mm^2 tensile strength and 7.4 x 10^{3}kg/mm^2 tensile elastic modulus. The fibre volume fraction is 35% and the weight fraction is 52%. The third is the composite reinforced by carbon cloth and is laminated with 50 plys by hand-lay-up method (hereafter called "Sample C"). The carbon cloth used for Sample C is "No 6343" supplied by Toray Industries Inc. and is woven from carbon bundles into a plain fabric. Each bundle has 3000 filaments which are 7 micron in diameter. The tensile strength and tensile elastic constant of this filament are 280 kg/mm^2 and 2.3 x 10^4 kg/mm^2, respectively. The fibre volume and weight fraction of Sample C are 48% and 56%, respectively.

For Samples B and C, the cured specimens are discs 130 mm in diameter and 12 mm in height. Since the mechanical properties of such materials are known to vary with off-axis angle,[7-9] test specimens are cut out with the off-axis angle of 0° to 45° in 15° increments so as to reveal the anisotropy. These samples are notched on their face and impacted in the laminated direction that is "flatwise". A cutting plan is shown in Fig 1.

The shape and dimensions of a sample are shown in Fig 2. Test details are as follows:

    beam length = 80 mm
    width = 4 $\pm$ 0.05 mm
    thickness = 10 $\pm$ 0.05 mm
    span = 60 mm
    notch: 45°-V-notch, tip radius = 0.25 mm
    depth = 1 $\pm$ 0.07 mm.

The notch was cut with a specially formed milling cutter.

2.2 Impact Test

The effect of striking velocity on impact properties is investigated on Sample A at room ($20 \pm 1^{\circ}C$) and liquid nitrogen temperatures. In the case of Sample B, the effect of the 'off-axis' angle (θ in Fig 1) is examined at both temperatures. In the former experiment, the impact velocity is varied by changing the initial position of the pendulum and calculated using the following expression;

$$v = (2gh)^{1/2}$$

    where v : the striking velocity
          g : the acceleration due to gravity
          h : the initial elevation of the pendulum.

When the test is made at liquid nitrogen temperature (LNT), the samples are immersed in liquid nitrogen for at least ten hours. After the samples fully attain LNT, they are quickly placed in position and tested. A spacer made from PVC (poly vinyl chloride) is placed on the anvil in order to adjust the impact position of the samples and to prevent rapid increase in temperature. Within 2.5 to 5.0 seconds specimens can be taken from the liquid nitrogen and the test completed. During this procedure the increase in temperature is considered to be not more than two degrees (K).

In general, for the Charpy impact test, a suitable machine capacity and impact velocity should be chosen according to the dynamic toughness of the test material together with its dimensions.

In the case of Sample A, the capacity of the impact machine is 10 kg-cm with typical impact velocity of 2.9 m/sec. For Sample B the capacity is 300 kg-cm and the impact speed is 4.56 m/sec. For Sample C the capacity and the impact velocity are 150 kg-cm and 3.35 m/sec, respectively. Five samples are tested under the same conditions. Correction for the energy loss caused by friction of the pendulum is carefully made in calculating the absorbed energy in all cases.

3. RESULTS AND DISCUSSION

3.1 Effects of Impact Velocity (Sample A)

Fig 3 shows the dependence of impact strength on impact velocity for Sample A at RT and LNT.

The open and closed circles represent the values at RT and LNT, respectively. Analysis of variance reveals that the impact strength depends upon temperature but not impact velocity. In other words, no dependence of impact strength upon impact velocity was observed at both room and liquid nitrogen temperatures within the present test condition, ie from 0.7 to 2.9 m/sec. The impact strength at liquid nitrogen temperature has a smaller value than that at RT. This phenomenon can be well recognised when the results are compared with the sectional views of fractured surface.

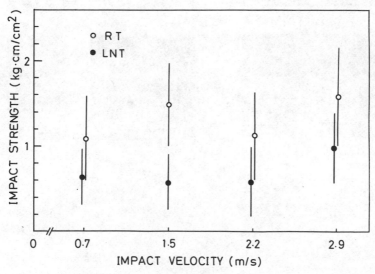

Figure 3.   Dependence of impact strength on impact velocity.

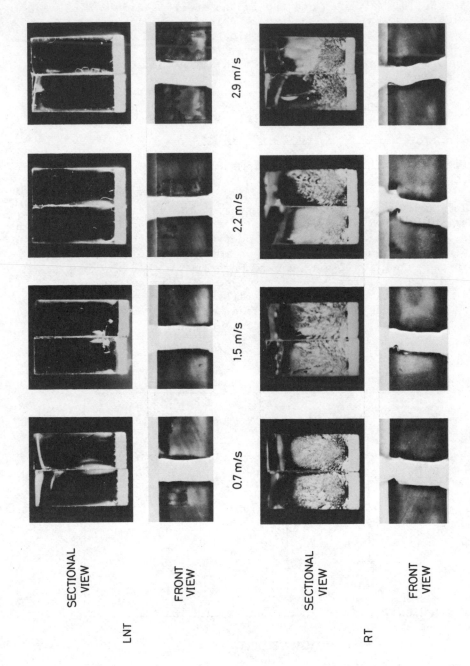

Figure 4.   Typical failure mode of fractured Charpy impact specimen.

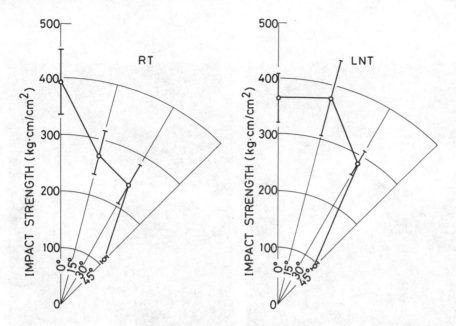

Figure 5. Effects of off-axis angle on impact strength of GRP.

Fig 4 shows the typical failure modes at RT and LNT, most of the surface area of the specimen tested at RT shows rough and large fibrous patterns and the front view shows the long crack path and consequently large fracture toughness. The surfaces of the specimen tested at LNT are flat and smooth, and the crack paths are straight showing a mode of brittle fracture. In the case of brittle fracture, the absorbed energy in Charpy impact test is approximately expressed as follows:

(absorbed energy) ≃ (fracture energy) + (projective energy).

Since the fraction of the projectile energy to the absorbed energy ranges from 1/3 to 1/2, that of the consumed energy to create the new surface, ie toughness to the total absorbed energy, is small.[10,11] Conversely, in the case of ductile fracture the projectile energy is small compared with the energy consumed in plastic deformation. Therefore, the dynamic toughness shows a decrease with temperature. Generally speaking, the changes in mechanical properties caused by decrease of temperature are apt to be considered beneficial, but materials should be carefully selected for cryogenic temperatures because of the decrease in impact strength.

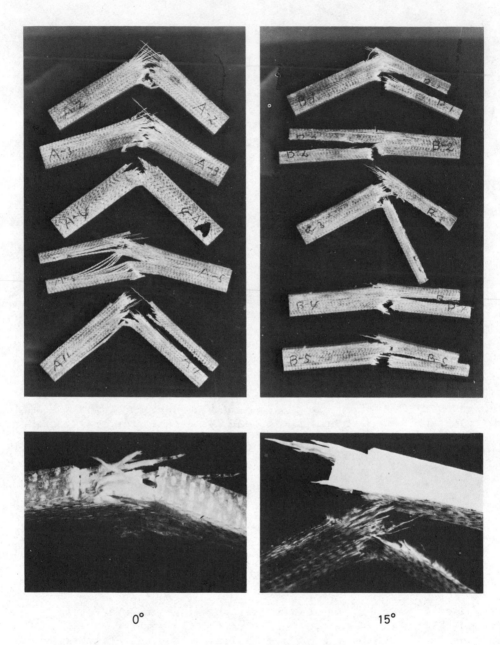

0°   15°

Figure 6.   Photographs of fractured GRP specimens at RT

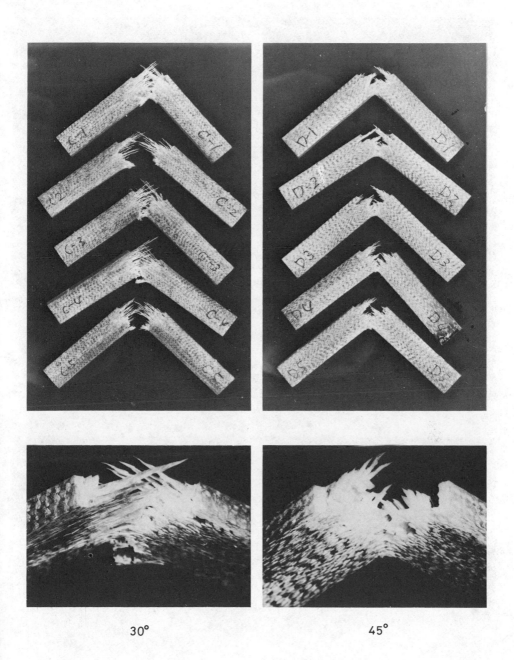

30° 45°

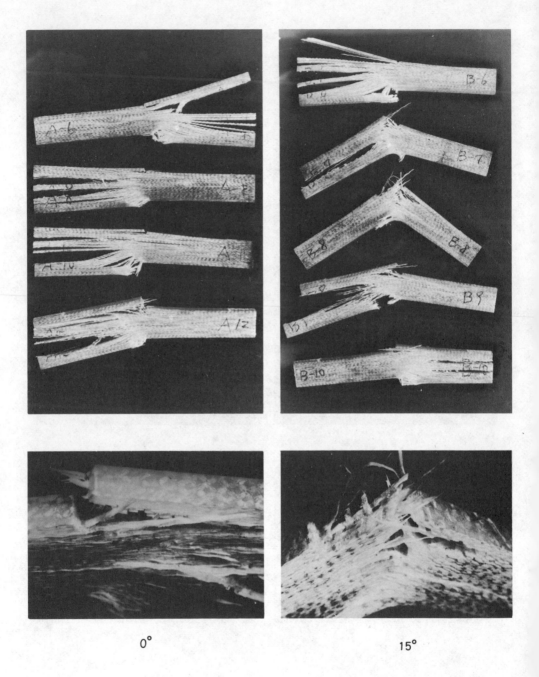

0°   15°

Figure 7.   Photographs of fractured GRP specimens at LNT

30° 45°

3.2 Effects of Off-axis Angle (Sample B)

Fig 5 shows the effects of off-axis angle on impact strength
at RT and LNT concerning Sample B (Glass Reinforced Plastic). In
this figure the radial axis and the angle show the value of impact
strength and off-axis angle of the specimen, respectively. At RT
the impact strength decreases with the increase of directional
angle and has the minimum value for 45° direction. Since the glass
cloth contained in Sample B is plain fabric and the longitudinal
strength of this cloth should be equal to the transverse strength,
the dependence of the impact strength on off-axis angle should be
symmetric with respect to the 45° direction.

The behaviour at LNT is similar to that at RT although the
highest value is not shown at 0° direction. The impact strengths
at LNT are slightly lower than that at RT, but higher at 15° direction. There are some reports[12] that the impact strengths of both
glass matt reinforced polyester and glass cloth reinforced polyester
increase with decrease of temperature from 90 to -80°C.[12] The
unexpected results in this work suggest that the impact strength of
FRP at low temperatures may be sufficient under certain conditions.

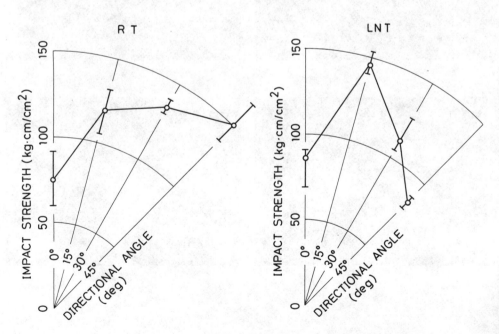

Figure 8. Effects of off-axis angle on impact strength of CRP

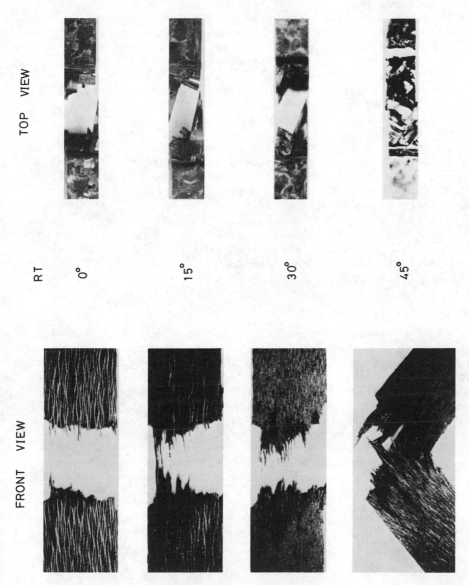

Figure 9. Photographs of fractured CRP specimens at RT

Figs 6 and 7 show photographs of fractured specimens at RT and LNT, respectively. They reveal that the fracture manner is the coexistent mode of fibre pull-out and fibre fracture accompanied by interlaminar peeling. If the fibre tensile strength is large compared to the shear strength between fibre and matrix, fibres will be pulled out. If shear strength is large compared to tensile strength then the fibre will fracture.[13] It is reported that the fracture of CRP (carbon reinforced plastics) and GRP mainly follow fibre fracture and pull-out, respectively, in failure.[14] The phenomenon of Sample B for which the angular increase makes a decrease in impact strength can be explained as follows. As GRP is thought mainly to fracture in the manner of fibre pull-out, the increase of the angle means a decrease in pull-out length of fibres and decrease of total shear stress to pull-out the fibres. Conversely, decrease of the angle causes an increase of pull-out length, which requires larger stress for pulling out the fibres. When the shear stress exceeds the tensile strength of the fibre, that is the off-axis angle is smaller than a certain critical angle, fibre fracture will occur. Because of this, the impact strength of 0° direction is not as large as the value expected from shear stress. It is expected, however, that the impact strength increases with the decrease of off-axis angle. This dependence of impact strength on off-axis angle coincides with that of the tensile strength of glass plain cloth reinforced plastics.[9]

The lower impact strength of 0° direction at LNT can be explained as follows. Fig 7 indicates that many interlaminar tears occur on the samples of 0° off-axis angle. These tears provide the ability of easy deformation on the specimen, because of this the tested specimens are not projected forward in the usual fracture mode, accompanied by apparent decrease of impact strength. This means that temperature decrease causes a decrease of impact adhesive strength of epoxide matrix between glass cloths. It is established that the interlaminar tears are dominant at high deformation rate in flexural test of GRP.[12] According to WLF (Williams – Landel – Ferry) theory, the increase of deformation rate is equal to the decrease of temperature and hence the proposition can be described as the interlaminar tears are dominant at low temperatures. This modified proposition appears to agree well with the behaviour of the specimens at LNT.

3.3 Effects of Off-axis Angle on Sample C

Effects of off-axis angle on impact strength of Sample C (CRP) at RT and LNT are presented in Fig 8. Radial axis and angle show impact strength and directional angle from fibre direction, respectively. The results at RT demonstrate that the impact strength increases with the directional angle and have the highest value at the directional angle of 45°. Though the behaviour at RT is different from that of GRP, the result at LNT is similar to that of GRP.

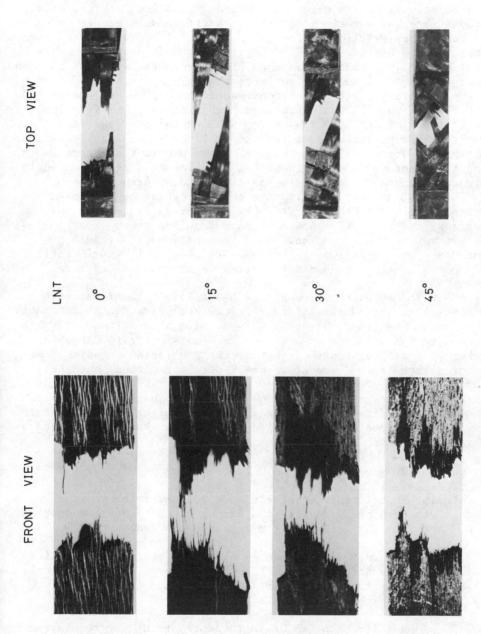

TOP VIEW

FRONT VIEW

LNT

0°

15°

30°

45°

Figure 10. Photographs of fractured CRP specimens at LNT

Fracture modes are represented in Figs 9 and 10. Figs 9 and 10 show fracture pattern of the specimens tested at RT and LNT, respectively. It appears that Sample C shows fibre fracture rather than fibre pull out as main failure mode resulting in brittleness. Carbon fibre has a serious demerit of brittleness[15] but is noted for its higher elastic modulus and specific strength compared with glass fibre. Because of this, the breaking mode of CRP is fibre fracture, while carbon fibres show a smaller shear strength between epoxy matrix and a smoother surface[9] than glass fibre. The increase of impact strength with directional angle at RT can be interpreted as follows. When the main fracture mechanism is fibre fracture the increase of angle decreases the applied force on one fibre resulting in the increase of the external force necessary for fibre fracture. Consequently, increase of angle provides an increase in impact strength. In the case at LNT, the impact adhesive strength between cloths and epoxy has a smaller value, as revealed by Sample B, and hence the pull-out mechanism cannot be neglected and causes the similar behaviour to GRP. Comparing Fig 9 with 10, the specimens tested at LNT show more marked pull-out behaviour than at RT. It may be concluded that the different impact strength dependence on off-axis angle between GRP and CRP is attributable to the two fracture mechanisms of fibre pull-out and fracture.

Even in GRP when the composite has a larger shear strength between fibre and matrix, fibre fracture mechanism can be dominant. In spite of its shear strength, GRP may also show a fibre fracture mechanism and has off-axis angle dependence similar to CRP when broken at small deformation, that is FRP fabricated edgewise[16] or having a large second moment of area. It is reported that the impact strength of GRP fabricated edgewise or having larger width,[12] increases with off-axis angle from 0 to 45°. The impact strength of the weakest direction must be the design criterion because three-dimensional forces will be applied in practice where GRP is used as a structural material.

4.   CONCLUSIONS

Charpy impact tests have been made on commercial epoxide resin, glass cloth reinforced epoxide and carbon cloth reinforced epoxide at LNT and RT. The following conclusions were drawn.

1)   Impact velocity dependence of impact strength of epoxide resin (Sample A) at both RT and LNT was not observed in this work. The impact strength at RT has larger values than those at LNT irrespective of impact velocity.

2)   Impact strength can be improved remarkably by reinforcement using glass or carbon cloths.

3)   Impact strength of reinforced plastics shows a dependence on off-axis angle and the dependence varies with the type of reinforcement, matrix and specimen shape.

4)   Impact strength of GRP is larger than that of CRP when the specimen shape is identical.

5.   ACKNOWLEDGEMENTS

The authors are grateful to Dr S Eumura and S Kobayashi of Government Industrial Research Institute Osaka for their help in the experiment. They would like to thank President Y Kobayashi of KOBAYASHIKIGATA Inc. for supplying specimens. They would also like to thank Dr T Hagihara of Osaka Kyoiku University for various stimulating suggestions. They wish to thank Mr T Nakamura of Ryoden Kasei Co Ltd for supplying the epoxide resin.

REFERENCES

1.   M Suzuki, S Iwamoto, Documents of the Committee of FRP
         Division, Society of Material Science of Japan (1978).
2.   S Nishijima, T Okada, Cryogenics 20 No 2 (1980) 86.
3.   K H Sayer, B Harris, J. Comp. Mat. 7 (1973) 129.
4.   E J McQuillen, L W Gause, J. Comp. Mat. 10 (1976) 79.
5.   Nippon Kaiji Kyokai, "AMMRC-Charpy V-notch Impact Test Method"
         Document of Committee No 28 (1973).
6.   K Okazaki, M Niguchi, K Shibayama, Koubunshi Ronbunshu Vol. 34
         No 3 (1977) 187.
7.   J M Lifshitz, J. Comp. Mat. 10 (1976) 92.
8.   D F Sims, J. Comp. Mat. 7 (1973) 124.
9.   T Hayashi ed. "Composite Material Engineering" Maruzen (1977).
10. S Uemura, Trans. Jap. Soci. Mech. Eng. 44 No 282 (1978) 1820.
11. S Uemura, Journ. of Ma-erial Testing of Mechanical Research
         Association (In Japanese) 23 No 4 (1978) 27.
12. Y Endo, H Yoshida, J. Soci. Mat. Sci. Japan 21 No 229 (1973)
         1765.
13. T Williams, G Allen, S M Kaufman, J. Mat. Sci. 8 (1978) 1765.
14. T Fujii, M Zaki, "Dynamics and Fracture in Composite Materials"
         Jikkyo Pub. Co. Ltd (1979).
15. S Kumazawa, J. Soci. Mat. Sci. Japan 21 No 229 (1972) 893.
16. S Uemura, Y Masuda, Bulletin of the Jap. Soci. Mech. Eng.
         No 738-1 (1973) 45.

NONMETALLIC AND COMPOSITE MATERIALS

AS SOLID SUPERLEAKS

J M Goldschvartz

Clavecimbellaan 273
2287 VK Rijswijk (ZH)
The Netherlands

INTRODUCTION

There are two types of superleaks, namely: the undesirable and
the desirable. The former corresponds to, for instance, a crack
in a Dewar containing He II and through which a superflow is
established, ie, it leaks the so-called superfluid component of
the He II. The latter is a device through which one needs, at
given circumstances, a flow of superfluid He II. Summarising, the
first is an accident and the second a device. This paper deals
only with the devices, in general solid porous materials in which
the so-called diameter of the pores, gaps, inter-crystalline spaces,
or small channels, etc, are equal or smaller than 100 Å. Of course,
the type of slits and capillaries used by Kapitza[1] and by Allen and
Misener[2] when they discovered the superfluidity are also superleaks.
However, the small channels they considered differed by approximately
two orders of magnitude to those considered in this paper.

Historically, the first superleak was recorded by Frederikse[3]
who used jeweller's rouge, (a fine powder about 0,04 μm), pressed
into a tube. This device was afterwards applied by Bots and Gorter[4]
as a superleak for their measurements of the fountain effect. This
material (Fe_2O_3) is still in use today as a superleak.

In 1956 Atkins et al[5] found that Vycor glass, an industrially
made porous glass, containing about 96% pure quartz, was a superleak with the properties shown in Table I, together with a slightly
modified form by Brewer, Champaney and Mendelssohn.[6]

Around 1965 the need for pure liquid ^{4}He was becoming urgent
and then the application of a superleak as a filter for macroscopic

277

Table I

Type of superleak	Onset temp. K	Author's diam. of the pores Å	Gap width from Ginzburg & Pitaevskii Å	Gap width from Mamaladze Å
Jeweller's rouge[4]	1.8	100	46	34
Porous Vycor glass[5,6]	1.64 2.0-2.1	50 70	38 68-106	26 56-100
SiC[7]	A 1.79 B 1.62	>70 <50	46 38	32 26
Wonderstone[11] half-hard natural	2.12 2.05	>70 >70	126 82	126 70
Talc-stone[15]	1.69	<70	40	28
Granite[16]	1.48	-	34	22
Lime-stone[16]	1.45	-	34	22
Magnetite[17]	A 1.57 B 1.52	- -	36 42	24 29

impurities became apparent and we started to use Vycor glass for that purpose. Moreover, the ^{3}He atoms were also kept apart from the superflow. Not only because they are fermions, as at that time was believed, but also because the temperature was too high. Nowadays we know that the superfluid ^{3}He exists with a transition temperature to phase A at 2.6 mK.

However the Vycor glass proved to be an unsatisfactory material: it is difficult to machine, breaks easily due to its brittle nature, lacks mechanical strength and often breaks under vacuum conditions in the cryostat. This is probably due to the freezing of the occluded water vapour and the sudden release of a large amount of occluded gases. A search for new solid materials to be used as superleaks started. Table I shows the features of the different types of superleaks discovered in the last 20 years or so.

Silicon Carbide[7]

We found that a dense form of reaction sintered silicon carbide[8] did behave as the desired superleak. This form of silicon carbide,

Fig 1. Photographs of liquid helium II drops at different stages, through a silicon carbide superleak.[7]

as well as others obtained by following different techniques, made by sublimation or by pyrolysis of silicon and carbon containing gaseous compounds,[8] can be produced in almost any desired porosity. Sintered silicon carbide porous plugs have already been used[10] as an element for controlling and measuring the gas flow into a vacuum system.

This form of reaction sintered silicon carbide is obtained by sintering technical-grade silicon carbide grains, smaller than 10 μm, impregnated by a carbonaceous substance, eg colloidal graphite, and heated in SiO vapour for several hours at $2000^{o}C$ in a hydrogen atmosphere. Porosity may be controlled by regulating the temperature, the reaction time and the repetition of the impregnation. Moreover, some important technical features make the silicon carbide particularly suitable for our purpose, namely: it is machinable, strong, chemically inert, thermally stable, may be directly sealed to Pyrex glass and can be cleaned and degassed at high temperatures. The samples used for our experiments, A and B in Table I, were porous plugs of approximately 9 - 10 mm diameter and 10 mm height, sealed in a Pyrex glass tube, vacuum tested and with a porosity of about 3% and 8%, respectively. In Figure 1 different stages of superfluid drops can be seen. These photographs were taken of sample B at a shutter speed of 10^{-3} sec and the direction of the movement of the curtain shutter was parallel and opposite to the direction of the movement of the drops.

Wonderstone[11]

Once we had the SiC superleak we continued our search for new natural materials, having in mind the porous volcanic lava type.

Results obtained with a natural material called 'wonderstone , which is a fine-grained sedimentary rock of volcanic origin found

in South Africa, showed that this material had some very desirable
features. Apart from the electrical, electronic, and chemical
applications of wonderstone, it is also used, after undergoing
different thermal treatments, to make cavities for electron spin
resonance spectrometers.[12] Wonderstone is an aluminium silicate[13]
with a porosity between 5% and 5.6% and its chemical composition
is: Silica 57.19%, Aluminium 32.78%, Ferric oxide 0.72%, Lime 0.40%
and Magnesia 0.36%. Wonderstone possesses some conspicuous characteristics. Table II details the properties of the material before
and after thermal processing. Prior to heating, wonderstone may be
cut easily, drilled, threaded, turned on a lathe, and polished.
The heating process resulted in a much harder material and also gave
rise to a significant increase in the physical dimensions of the
sample. All samples used for our experiments were cut at random
from a large piece and fixed to a Pyrex cell by means of aluminium
flanges. The lambda tightness was obtained with indium wire covered
with some silicon vacuum grease.[14] The onset temperatures of the
natural and half-hard samples are 2.05 K and 2.2 K, respectively.
The pores of the hard sample were so large that it was leaking
He I. The cell and the manner in which the superleaks are fixed
may be seen in Figure 3.

Talc-stone[15]

Another natural material which is a good solid superleak is
talc-stone. It is a magnesium silicate from Rabenwald, in Austria,
with the chemical composition: SiO_2 - 51.6%; MgO - 30.9%; H_2O - 4.7%;
CO_2 - 2.7%; Al_2O_3 - 8.8%; FeO - 1.0% and CaO - 0.3%. It has a
crystalline structure, occurs naturally in blocks, is light grey,
almost white in colour, with a notable greasy touch. Talc-stone
may also be machined and shaped easily but it is mechanically
inferior to wonderstone. The sample used for our experiments to

Table II

Sample	Original dimens.		Final dimens.		Variation		Total time of thermal process (hours)	Max. temp. $^\circ$C	Final co-lour
	Ø mm	Thickness mm	Ø mm	Thickness mm	Ø %	Thickness mm			
Natural	35	5	35	5	–	–	–	–	grey
Half-hard	35	5	35.9	–	2.5	–	24	750	grey-pink
Hard	35	5	35.2	5.1	3.4	2	72	1200	pink

determine the onset temperature, was a disc of 40 mm diameter and
10 mm thickness, cut at random from a natural block and fixed to
the cell in the same way as the former material. The onset temperature was found to be 1.69 K.

Rocks as Superleaks[16]

We found that two types of simple rocks taken from different
mountains in Europe had the properties of fine superleaks. These
rocks are: a fine grained grey granite-stone from Poland and a
lime-stone from Urach, Schwäbisch Alb, West Germany. The granite-
stone is a holocrystalline rock, with grains of medium size and the
approximate mineral composition: plagioclase 39%, quartz 27%,
alcaly feldspar 23%, biotite 9%, and accessory minerals 1%. As far
as the lime-stone is concerned, it is of the type known as 'Platten
Kalk', comparable to lithographic lime-stone. It is calcilutite,
an extremely fine grained rock, made up of about 96% to 98% of
calcium carbonate with impurities of magnesium carbonate, clay,
iron oxide, traces of manganese, and small quantities of quartz.

Using polished samples of diameter 40 and 25 mm and thickness
10 and 7.5 mm for the granite and the lime-stone, respectively, the
onset temperature for the superflow through both samples was determined and found to be 1.48 K and 1.45 K for the granite and for the
lime-stone, respectively.

Magnetic Superleaks[17]

Two new natural materials which behave as superleaks to superfluid ^{4}He and also have magnetic properties have been found. In
the search for magnetic superleaks the first choice was, obviously,
a highly magnetic material such as manufactured ferrite which has
the advantage that its porosity is uniform. We tested such a
material but apparently not at a low enough temperature to be able
to detect any onset point for the superflow of ^{4}He. The pores or
gap distances must be extremely small. Second, natural magnetic
materials, among which the most obvious were different types of
the so-called magnetites, were tested. Two magnetites of apparently
quite different origin have been tested with superfluid ^{4}He. One
from the south-east of Argentina (sample A) and another of unknown
origin taken at random from the Mining Department of the Delft
University of Technology, The Netherlands (sample B). An X-ray
diffraction analysis of these two samples gave the following
components: for sample A, heamatite (Fe_2O_3) (ferromagnetic),
magnetite (Fe_3O_4) (ferromagnetic), mica and garnet $(Mg,Fe)_3Al_2$
$(SiO_4)_3)$ (paramagnetic) and for sample B, mainly magnetite (Fe_3O_4)
(ferromagnetic) and chlorite, a sheet silicate $(((Mg,Fe)_5AlSi_3O_{16})$
$(OH)_8)$ (paramagnetic). Both samples are anisotropic aggregates
and the magnetic characteristics are given at room temperature.
The obtained onset temperatures are: for sample A - 1.57 K and for
sample B - 1.72 K.

The search for magnetic superleaks had its origin in the anisotropic magnetic superfluid phases of liquid ^{3}He. The application
of magnetic superleaks to the peculiar anisotropic magnetic superfluid phases of liquid ^{3}He at very low temperatures[18,19] might be
relevant although not an easy problem. In that physical situation
those phases would be influenced by an adequately magnetised finite
geometry medium, not necessarily those described here, which they
filled, since below the transition temperatures the liquid has the
properties of a superfluid and magnetic properties, similar to those
of nematic liquid crystals.

The Onset Point of a Superleak

From the experiments performed by Atkins et al[5] it is known
that in the thermomechanical effect, the superfluid does not start
to flow into a narrow channel of a superleak at the lambda transition
but only at some lower temperature. This phenomenon is very similar
to the one obtained with unsaturated helium films in which the onset
of the flow is also below the normal lambda point and was studied
some years ago by Bowers et al[20] and by Long and Meyer.[21,22] These
temperatures are called the 'onset temperatures' and the fact that
they are below the normal lambda transition is called 'depression
of the onset of the superflow' or 'shift of the lambda transition'.
In order to estimate the widths of the gaps or the so-called diameters of the pores through which the superfluid penetrates, we
used the relationships between that depression of the transition
temperatures and the width of those pores or gaps. This relationship, derived for the parameters of the phenomenological theory of
superfluidity for thin films of He II and its behaviour in narrow
gaps, is variously written as:-

$$\Delta T_\lambda = -\frac{2 \times 10^{-14}}{d^2}\ K \quad \text{and} \quad \Delta T_\lambda = -\frac{2.5 \times 10^{-11}}{d^{3/2}}\ K,$$

where ΔT_λ is the shift of the lambda point and d is the thickness
of the film measured in centimetres. The first relationship was
derived by Ginzburg and Pitaevskii[23] and the second, which seems to
be more consistent with the existing experimental data, is due to
Mamaladze.[24] It was, however, pointed out[25] that the width of a
parallel channel, which probably occurs most frequently, is 2d.
Using this latter criterion and the two expressions given above,
the widths of the channels shown in Table I were estimated.

The smaller the pores, the lower the onset temperatures. The
channel connections which run between both sides of a superleak are
not uniform and interconnect at random. If we call $\emptyset_{min}$ the
minimum diameter along a connecting path, then the largest $\emptyset_{min}$
of all possible paths determines the onset temperature; it is in
that one that the movement of the superfluid will take place first,
ie, at the highest temperature below the lambda point. This is

called a size effect[26] and results from interactions between the
excitations of the quasi-particles of the liquid helium two and the
solid. When the temperature is high and the gap (or diameter)
small, the interaction is equivalent to a viscosity effect. The
number of excitations diminishes with the temperature.

However, it is interesting to note[16] that in Ginzburg and
Pitaevskii's equation d is the thickness of the superfluid film
while in Mamaladze's equation d is the gap width. Nevertheless,
in Mamaladze's as well as in a paper by Gamtselidze, Dzhaparidze
and Turkadze,[27] the results obtained using the two equations are
compared without any comment.

Determination of the Onset Point

Atkins et al[5] defined the onset temperature of a superleak as
the temperature at which the level of the liquid He inside a cell
moved at about 0.1 mm S^{-1}. This method is dependent on the visual
movement of a liquid column and also on the measurement of time,
but any method that eliminated the possibility of human error is
likely to be more accurate and would be preferred. However, all
methods have an element of uncertainty due to the fact[28] that there
is a flow of saturated vapour of ^{4}He and a capillary condensation
above as well as below the lambda point.

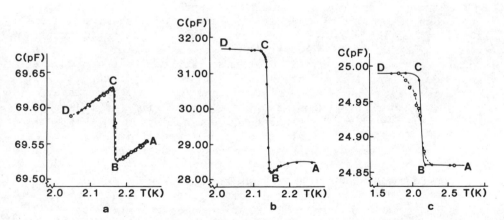

Fig 2.   Measurements of the capacitances of plane-circular capacitors
with solid superleaks as dielectric.[29]
a. Dielectric: Wonderstone. Two series of measurements
made with the same capacitor.
b. Dielectric: Talc-stone.
c. Dielectric: Vycor glass. Full line represents the cooldown process and the dashed line was produced on warming.

Table III

Materials* Figure 2	Known values at room temperature		Calculated values from Figure 2	
	Dielectric constant	Porosity (%)	Dielectric constant	Porosity (%)
(a) Wonderstone	21.3	5 to 6	20	10
(c) Vycor glass	3.8	28	3	25

* The values of the talc-stone at room temperature were unknown.

However, it was not entirely clear whether the submicroscopic pores (< 100 Å) of a solid superleak were filled with liquid helium above the lambda transition, or even above the onset temperature. To resolve this uncertainty, the capacitance of plane-circular capacitors with solid superleaks as dielectrics were immersed in liquid helium and measured.[29] The results of these measurements are shown in Figure 2, where it is clear that the pores are free from liquid helium above the onset temperature. This method also indicates the onset temperature, which is the temperature at which the rapid change in the values of the capacitance takes place, point B, and presents the possibility of measuring the filling of the dielectric-superleak, segment BC, not attempted in this work. The values in Table III, calculated from the data in Figure 2, provide a quantitative basis for the results obtained.

Considering a superleak as a device used in low temperature technology to fill or to empty a cell inside the liquid helium bath, the method[11] shown in Figure 3 produces acceptable results. The value of the resistor R changes dramatically when liquid helium two is present inside the cell. This apparatus can also be used with Atkin's method since it is provided with a calibrated capillary tube in the upper part. An improvement in the accuracy of this method might be possible by painting a resistor of colloidal carbon or by evaporating a metallic superconductive resistor onto the inner surface of the superleak. The superconductive material could

be an alloy with a transition temperature as near as possible to
the lambda point. The transition temperature from the normal to
the superconducting state could be adjusted by an external magnetic
field.

In this kind of measurements there is always a real doubt that
the results apply to a true superleak.[17] For instance, the manner
in which the superleak is fixed to the system, whether it is soldered
to the glass of the cell or fixed with an indium wire gasket, might
represent a microscopic leak. Further, it is impossible to apply
any kind of reliable leak detector to a superleak. It is, however,
possible to overcome this doubt by making a reasonable number of
observations and, as far as possible, mounting and dismounting the
same sample and observing the reproducibility of the measurements.

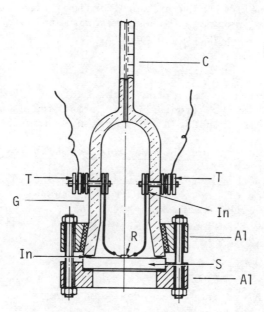

Fig 3.   Bottom of the cell used for the determination of the onset
         temperature of solid superleaks.[11]

         C - Pyrex glass calibrated capillary;
         G - Pyrex glass flange;
         T - electrical terminals;
         In - indium seals;
         Al - aluminium flanges;
         S - superleak disk;
         R - detector resistor.

For the determination of the onset temperature, a good method, although rather difficult to apply, was derived from experiments made to measure the 'stopping power' of a solid superleak, ie, measurements of the power necessary to warm the superleak to such an extent as to stop the superflow.[30] The results of these experiments are shown in Figure 4 for different sizes of wonderstone superleaks. The stopping power turned out to be a linear function of the temperature and, as expected, the straight lines intersect the temperature axis at the onset temperature. The mean value of these intersections, obtained by a computer least-square analysis for each superleak, or for each line, is 2.10 K. Thus, the original value, determined using the method shown in Figure 3, was within 2% of this true value.

SOME APPLICATIONS OF SUPERLEAKS

a. **As a Filter**

To fill a cell with pure liquid helium, free from macroscopic impurities. For instance, for the experimental study of the electrical breakdown of liquid helium.[31]

b. **As an Isotope Separator**

This is obvious because the light isotope ^{3}He is not carried by the superflow of the liquid ^{4}He at temperatures above the transition temperature to phase A of superfluid ^{3}He (2.6 mK).

c. **As a Separator in the ^{3}He-^{4}He Dilution Refrigerator**

Strictly speaking, this application, proposed by London[32] in his original idea of the refrigerator, is the same as the former. It is, however, used in the construction of the ^{3}He-^{4}He dilution refrigerator, to separate the liquid ^{3}He from the liquid superfluid ^{4}He for the circulation of the former.

d. **In a Vortex Refrigerator**

Staas and Severijns[33] have designed a refrigerator based on the interaction between the superfluid and the normal fluid flow through a capillary. The superleak is used to 'separate' the normal from the superfluid liquid.

e. **In a Servo-valve for Liquid Helium Two: The Cocatron**

This represents the simultaneous application of both superfluidity and superconductivity.[34] In Figure 5 an outline of the Cocatron is shown. In the cell C, placed in the liquid helium bath in Cryostat D, a constant height h of liquid helium two is maintained. At this level a 'superconductive resistor' R is set up. The bottom

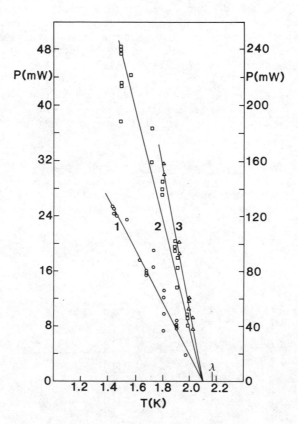

Fig 4.   Stopping power as a function of the temperature.[30] Lines 1
         and 2 are referred to the left hand side ordinates and the
         line 3 to the right hand side ordinates axis.

of the cell is closed by means of a flat solid superleak S provided
with a heater H_1. A carbon resistor T allows the temperature of
the liquid helium bath to be controlled.

When the temperature of the helium bath reaches the onset
temperature of the superleak, the superfluid flows through the
superleak due to the thermo-mechanical effect. The heat required
to produce this effect is provided by the heater H_2. When the
liquid helium two inside the cell reaches the superconductive
resistor R, it becomes superconducting and then, through the electronic switch, the heater H_1 of the superleak S raises the local
temperature above the onset temperature. Thus, the flow of the
superfluid through the superleaks is halted. When the liquid
helium two in the cell evaporates, the level h falls below that of
the resistor R. The latter becomes normal and consequently the
heater H_1 is prevented from warming the superleak. Therefore, the
superfluid starts to flow again filling the cell, and the cycle is
repeated.

f. Method of Measuring the Size of Sub-microscopic Pores

Applying the equations given above, the size of sub-microscopic
pores could be estimated by the determination of the onset tempera-

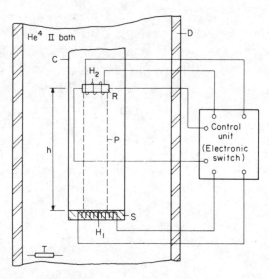

Fig 5. Outline of the Cocatron, a servo valve to keep a constant
       level of liquid helium two.[34]

tures, provided one has the apparatus to obtain sufficiently low
temperatures. With small pores as in some natural materials, it is
necessary to cool the liquid helium four to temperatures only
obtained with liquid helium three cryostats or even with ^{3}He-^{4}He
dilution refrigerators.

g. Ultra Cold Neutrons (UCN)

Neutrons with very low energy, ie, ultra cold neutrons (UCN),
can be stored concentrated and, due to their long life, permit the
measurement of fundamental constants.[35,36] Golub and Pendlebury,[37]
on the one hand, have suggested the use of liquid ^{4}He at a temperature of about 1 K to produce UCN. On the other hand, the neutron
absorbent isotopic impurities of ^{3}He in the normal commercial liquid
^{4}He would be an obstacle in obtaining the desired concentration of .
UCN. Scott and McClintock[38] commented that the separation technique
of Fatouros et al[39] using the method (b) mentioned above, and Vycor
glass, gave insufficient pure ^{4}He because the ratio obtained
between the two isotopes was still an order of magnitude too small
for the production of UCN. Atkins and McClintock,[40] again using
Vycor glass but refining the technique 'applying the phenomenon of
heat flush to remove the ^{3}He atoms from the vicinity of the superleak', claim they have achieved the necessary purity, with an
isotopic ratio $R_{4,3} = 10^{12}$. However, changing the type of superleak
could improve this figure, ie, using a superleak with smaller pores
and, consequently, lower temperatures. Given that the transition
temperature to superfluid ^{3}He is about 10^3 lower than the transition
temperature of ^{4}He, this means that the gaps of the superleak could
be very much smaller and the temperature still be kept well above
the transition temperature of ^{3}He.

h. Superconductors Pressed into Porous Materials

This is not precisely an application of a superleak. Nevertheless, it is mentioned as a 'use' of a non-metallic porous
material in low temperature technology. In fact, Watson[41,42,43]
found that indium (In) as well as an alloy of lead (Pb) and bismuth
(Bi) pressed into the pores (< 100 Å) of a glass produced remarkable
effects. The transition temperature of the indium turned out to
change with the diameter of the pores, increasing with decreasing
pore size; the critical magnetic field also changed with the pore
size. However, it must be stated that no other superconductor
pressed in a non-metallic material such as wonderstone or talc-
stone, have been tested.

ACKNOWLEDGEMENT

This paper, especially prepared for the ICMC, is fundamentally,
a resume of a series of previously published papers by the author.

REFERENCES

1.   P L Kapitza, J. Phys. Moscow, 5 (1941) 59.
2.   J F Allen and A D Misener, Proc. R. Soc., A172 (1939) 467.
3.   H P R Frederikse, Physica, 15 (1949) 860.
4.   G J C Bots and C J Gorter, Physica, 22 (1956) 503.
5.   K R Atkins, H Seki and E O Condon, Phys. Rev., 102 (1956) 582.
6.   D F Brewer, D C Champaney and K Mendelssohn, Cryogenics, 1
         (1960) 108.
7.   J M Goldschvartz and B S Blaisse, Proc. ICEC 2 (1968) 304.
8.   Developed and provided by Philips Research, Eindhoven, The
         Netherlands.
9.   W F Knipenberg, Philips Res. Rep., 18 (1963) 161.
10.   R G Christian and J H Leck, J. Scient. Instrum., 43 (1966) 229.
11.   J M Goldschvartz, E Martin and B S Blaisse, Cryogenics, 10
         (1970) 160.
12.   M J Bakker and J Smidt, Appl. Scient. Research, B9 (1962) 199.
13.   L T Nel, H Jacobs, J T Allen and G R Bozzali, Geological Series
         Bulletin, No 8, Union of South Africa, Dept. of Mines (1937).
14.   J E Vos and R Kingma, Cryogenics, 7 (1969) 50.
15.   J M Goldschvartz and B S Blaisse, Supp. Bulletin Intern. Inst.
         Refrig., Commission I, Tokyo (1970) 231.
16.   J M Goldschvartz, A Kollen, F Mathu and B S Blaisse, Cryogenics,
         13 (1973) 303.
17.   J M Goldschvartz, Nature, 266 (1977) 824.
18.   A J Legget, Rev. Mod. Phys., 47 (1975) 331.
19.   J C Wheatley, Rev. Mod. Phys., 47 (1975) 415.
20.   R Bowers, D F Brewer and K Mendelssohn, Phil. Mag., 42 (1951)
         1445.
21.   E S Long and L Meyer, Phys. Rev., 85 (1952) 1030.
22.   E S Long and L Meyer, Phys. Rev., 98 (1955) 1616.
23.   V L Ginzburg and L P Pitaevskii, Sov. Phys. Jetp., 7 (1958) 858.
24.   Yu. G. Mamaladze, Sov. Phys. Jetp., 25 (1967) 479.
25.   A J Symonds, D F Brewer and A L Thomson, Quantum Fluids,
         D F Brewer (Edit.), North Holland, Amsterdam (1966) 267.
26.   M Kriss and I Rudnik, J. Low Temp. Phys., 3 (1970) 339.
27.   G A Gamtselidze, Sh. A. Dzhaparidze and K A Turkadze, Sov.
         Phys. Jetp Lett., 6 (1967) 44.
28.   E F Hammel and A F Schuch, Proc. Low Temp. Phys. and Chem.,
         Wisconsin (1958) 23.
29.   J M Goldschvartz, W P Van der Merwe, E Bodegom and J Dam,
         Cryogenics, 19 (1979) 679.
30.   J M Goldschvartz and J H M Steman, accepted to be published in
         Cryogenics later this year (1980).
31.   J M Goldschvartz and B S Blaisse, Cryogenics, 6 (1965) 169.
32.   H London, Proc. Low Temp. Phys. Conf., Oxford (1951) 157.
33.   F A Staas and A P Severijns, Cryogenics, 9 (1969) 422.
34.   J M Goldschvartz and J G Krom, Cryogenics, 17 (1977) 577.
35.   P V E McClintock, Nature, 275 (1978) 174.
36.   I I Purica, Rev. Roum. Phys., 24 (1979) 243.

37.   R Golub and J M Pendlebury, Phys. Lett., 62A (1977) 337.
38.   R J Scott and P V M McClintock, Phys. Lett., 64A (1977) 205.
39.   P P Fatouros, D O Edwards, F M Gasparini and S Y Shen, Cryo-
      genics, 15 (1975) 147.
40.   M Atkins and P V E McClintock, Cryogenics, 16 (1976) 733.
41.   J H P Watson, Phys. Rev., 148 (1966) 223.
42.   J H P Watson, Phys. Lett., 25A (1967) 326.
43.   J H P Watson, Phys. Rev. B, 5 (1972) 879.

LOW TEMPERATURE PROPERTIES OF CARBON FIBRE

REINFORCED EPOXIDE RESINS

W Weiss

Messerschmitt-Bolkow-Blohm GmbH
Ottobrunn, Germany

INTRODUCTION

In recent years Carbon Fibre Reinforced Epoxide resins have
been used for applications in the aerospace industry. The high
ratios of strength and stiffness to specific weight are important
in structures such as satellites and space vehicles. For
applications at room temperature the material properties were also
studied with care. The test equipment, the specimens and the manufacturing procedures were controlled at a high standard. A high
tensile or a high modulus carbon fibre is used according to the
particular requirement. As many structures will work at low temperatures in the future, the material properties at these temperatures
are required for design purposes. The most important properties are
strength, stiffness and thermal characteristics.

SELECTION OF MATERIALS AND MAIN PROPERTIES

Some characteristics of fibres and resins vary considerably.
For example the Young's modulus of a typical high tensile fibre and
a high modulus fibre are 220 Gpa and 370 GPa, respectively.
It was therefore intended to test materials with extreme properties.
The fibres chosen were the high tensile fibre T300 and the high
modulus fibre M40A, both produced by the Torayca company.

Two expoxide resin systems were selected for examination. One
resin was chosen because of its reported low temperature characteristics, the other should have acceptable properties at room temperature and below. The semiflexible epoxy CY221/HY979 has been
reported as a suitable material for use in superconducting magnets
(1). The rigid epoxy LY556/HY917 was suggested by the producer

MECHANICAL PROPERTIES			THERMAL PROPERTIES
STATIC PROPERTIES		FATIGUE PROPERTIES	
STIFFNESS PROPERTIES	STRENGTH PROPERTIES		
$E\|_t$ $E\perp_t$	$\sigma\|_{tF}$ $\sigma\perp_{tF}$	S–N curves for pulsating and alternating stresses for tension, flexure and shear	$\alpha_t\|$ $\alpha_t\perp$
$E\|_t$ $E\perp_t$	$\varepsilon\|_{tF}$ $\varepsilon\perp_{tF}$		$\lambda\|$ $\lambda\perp$
$G\#_t$ $V\|\perp_t$	$\sigma\|_{cF}$ $\sigma\perp_{cF}$		C_p
	$\varepsilon\|_{cF}$ $\varepsilon\perp_{cF}$		
	τ_{ILSS}		

FIGURE 1 MATERIAL PROPERTIES TO BE TESTED

(Ciba-Geigy Ltd) as a suitable resin for the whole temperature
range.

Because of the large number of tests it was decided to measure
two of the systems with a good statistical base.

For the design of structural parts a large number of different
laminates can be used. Therefore the number of unknown material
values is also large. If however the properties of the unidirectional laminate (UD) are known, the properties of any laminate can be
calculated (2). Each laminate can be split up theoretically into
UD-laminates. The characteristics of these are the stiffness values
$E||$, $E\perp$, $G\#$ and the Poisson-ratio $\nu||\perp$, the strength values $\sigma||_F$
and $\sigma\perp_F$, the ultimate strains $\varepsilon||_F$ and $\varepsilon\perp_F$ for tensile and compressive
load, the dynamic strength (fatigue) by pulsating and alternating
load in tension, flexure and shear and the thermal properties $\alpha t||$,
$\alpha t\perp$, $\lambda||$, $\lambda\perp$ and Cp (Fig.1). In addition to the stiffness values
the thermal properties of any laminate on the basis of the charac-
teristics of the UD-laminate (depending on the fibre content) may
also be calculated with the help of the superposition theorem and
micromechanic models (Fig.2).

EVALUATION AND STRESS ANALYSIS OF DIFFERENT SPECIMENS

At room temperature the tension samples are held with grips,
which are pressed against the sample by a mechanical or hydraulic
device. Four loading pads are glued to the sample to prevent local
damage. At low temperatures the composites contract more than the
steel grips and compensation for shrinkage with an hydraulic device
is not possible. Using mechanical grips there is the possibility
that the sample will slip from the grips. Two methods were tried
to solve the problem:

- Samples with wedge-shaped loading pads (Fig. 3)
- Samples in grips, which are pressed by screws (Fig.4)

A force is introduced into the wedge-sample from the grips and,
this sample did not work in practise. Shear failure was induced in
the wedges the sample itself did not fracture. The measurement of
the stiffness was inaccurate as a result of bending moments induced
by different wedges. The clamped samples work well at low temperatures (Fig.5). The samples fracture in the mid region and the
stress-strain diagram give reasonable values.

Waisted samples are used to measure the tensile and compressive
strength (Fig.9). The waisting is necessary to prevent fracture
at the end of the loading pads. The tensile sample was idealized
by finite element analysis to ensure that the stresses in the middle
of the sample are higher than in any other part due to the waisting

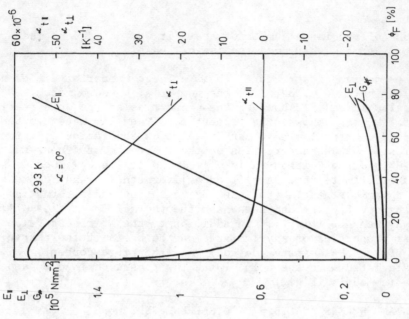

Figure 2b. Calculated Stiffness and Thermal Expansion Properties derived from Unidirectional Laminate Values.

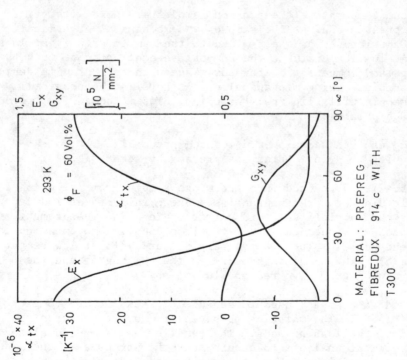

Figure 2a. Calculated Stiffness and Thermal Expansion Properties dericed from Unidirectional Laminate Values.

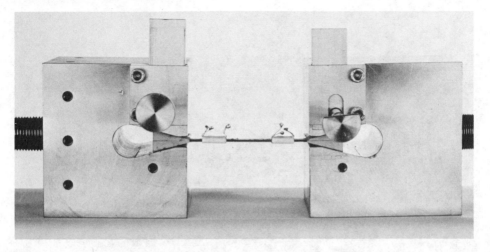

Figure 3.   Tensile Test Specimen with Wedges

(Fig.10). The shear stresses are small, so that there will be no premature shear damage. A parametric study of the gripping force F_Y showed that it has no relevant influence on the shear and longitudinal stresses. All mechanical values parallel to the fibres can be achieved with this one specimen. The waisting is not necessary for the determination of the properties perpendicular to the fibre. The shear modulus G was measured with an unwaisted tensile sample consisting of a $\pm 45^\circ$ balanced laminate. It may be calculated from the tensile stress σ_x, the tensile strain ε_x and the strain perpendicular to it ε_y by the following formula:

$$G = \frac{\sigma_x}{2(\varepsilon_x - \varepsilon_y)}$$

The compression sample (Fig.9) was also idealized and the stresses were claculated (Fig.11). In order to prevent buckling, the distance between the pads should be small. Therefore no straight part was provided in the middle of the sample. The stresses in the cross-section of the centre are not constant as a result of the notch.   The stress at the surface of the root of the notch is 32% higher than the mean stress. The stresses due to the grips have no influence on the damage. The shear stresses are high and may cause shear damage, if there is poor bonding between the pads and the sample.

The compression forces are applied to the sample by cones. These cones do not work at low temperature, because there is no lubrication possible. No steady stress-strain curve could be

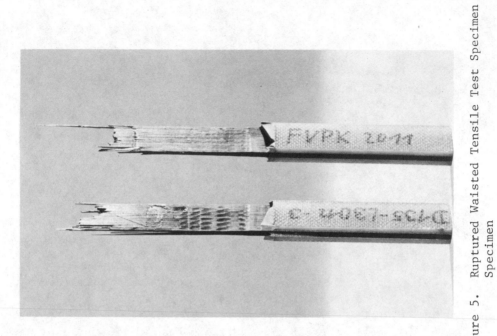

Figure 5. Ruptured Waisted Tensile Test Specimen
           Specimen

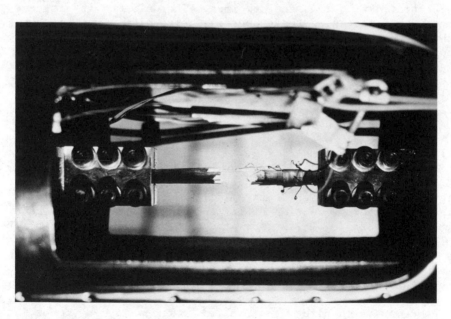

Figure 4. Holding Device for Tensile Test
           Specimen

obtained in preliminary tests. Further studies to eliminate this effect are necessary.

The interlaminar shear strength was determined with a short-flexure sample with the dimensions 2 x 10 x 16 mm.

The thermal properties were measured with tubes, which have the same dimensions for all the different measuring devices. This allows rational manufacturing of the samples.

SAMPLE MANUFACTURE

The production of samples out of fibre reinforced composites can be divided into five steps. The first is to wind the laminate, if a prepreg is not available. The second is to lay the laminate on a plate or wind it around a tube with the fibres in the desired directions. The next step is to cure the laminate under pressure and increased temperature in an autoclave. Then the loading pads are glued to the plate, if the samples are provided for mechanical testing. The last step is machining the plate, e.g. waisting by grinding and cutting the plate for tensile test specimen.

TEST EQUIPMENT FOR LOW TEMPERATURES

In the test machines the sample is surrounded by the cooling liquid (Fig.6). The load from the electrodynamic drive of the 100 kN machine for fatigue measurements and the spindle drive for static measurements is carried to the sample by two tubes. The inner tube is pulled and the outer tube is compressed for a tensile load. To change the sample, the cryostat is moved downwards with a screw elevator. The load cell is positioned in the middle of the pull rod outside the cryostat. The large openings in the outer tube allow a quick change of the sample and provide good maintenance facilities (Fig.7), with the bellows ensuring an adequate seal at the cryostat. The steel sample is inserted for investigating the frequency range of the machine.

In contrast to the 100 kN mechanical test machine, the 16 kN machine has the load cell near the sample in the cold region (Fig.8). The sample for tension tests is held by grips. The cryostat with the vacuum gauge tubes is in the down position. This machine produces the force by means of an hydraulic drive system.

The thermal expansion coefficient was measured with a laser dilatometer, where the expansion was expected to be small. Where the dilation was large, controlling measurments can be performed with an inductive dilatometer.

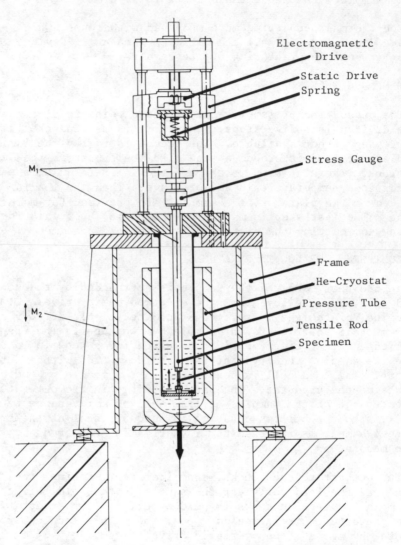

Figure 6.   Design of Fatigue Testing Machine

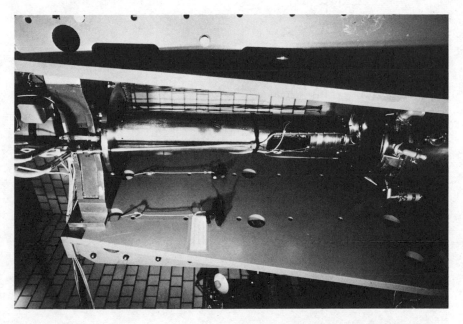

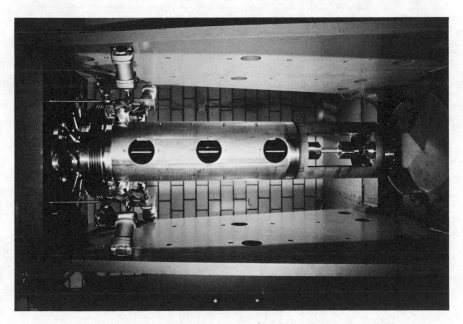

Figure 7. Fatigue Testing Machine with Electro-dynamic Drive

Figure 8. Fatigue Testing Machine with Hydraulic Drive

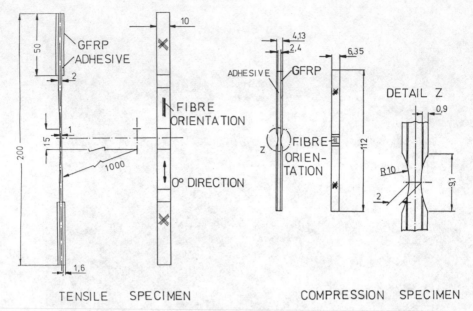

TENSILE   SPECIMEN   COMPRESSION   SPECIMEN

Figure 9.   Waisted Tensile and Compression Specimen

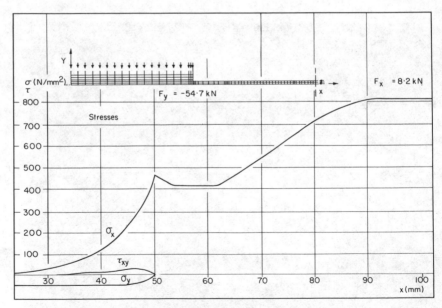

Figure 10. Idealisation of the Tensile Specimen in Finite
Elements and resulting Stresses

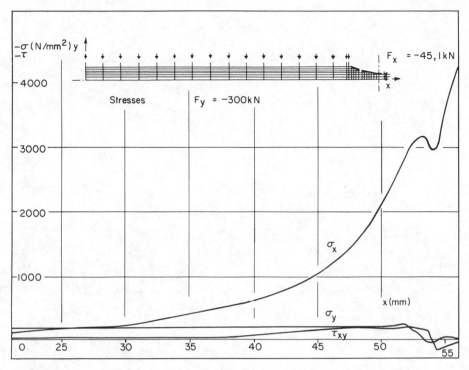

Figure 11. Idealisation of the Compression Specimen in Finite Elements and resulting Stresses

MECHANICAL PROPERTIES							
RESIN	FIBRE	TEMP [K]	FIBRE ORIENTATION	E [kNmm^{-2}]	ν [-]	σ_F [Nmm^{-2}]	ε_F [%]
CY 221/ HY 979	T300	293	‖	132	(0,34)	1700	1,22
		77	‖	141	0,32	2010	1,34
	M 40A	77	⊥	11,45	0,012	42,2	0,37
LY 556/ HY 979	T300	293	‖	135	0,31		
		77	‖	137	0,31		

Figure 12. Mechanical Properties of Unidirectional Laminates at 293 K and 77 K

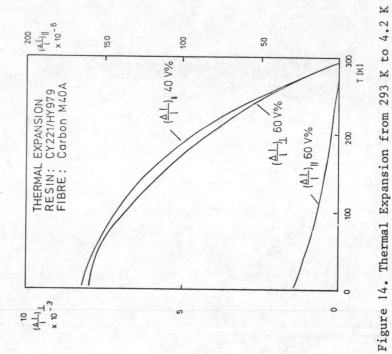

Figure 14. Thermal Expansion from 293 K to 4.2 K

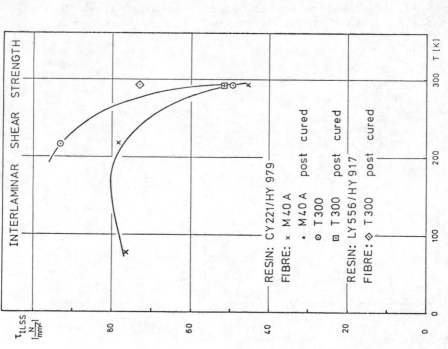

Figure 13. Interlaminar Shear Strength at 293 K, 218 K and 77 K

Figure 15. Chain of GFRP-Lugs in the testing device

RESULTS OF MECHANICAL AND THERMAL TESTS

First test results, which can be reported, are on mechanical and
thermal properties. The mechanical properties measured are the Youngs
modulus $E||$ the Poisson ratio ν, the fracture strain ε_F, the fracture
stress σ_F and the interlaminar shear strength τ_{ILSS}. The Young's
modulus parallel to the fibres of the composite with the semiflexible resin (CY221/HY979 with the high tensile fibre M40A), rises
from 132 GPa at room temperature to 141 GPa at 77K, which means an
increase of 7% (Fig.12). With the rigid resin (LY556/HY917) the rise
is only 2%. Poisson's ratio of the semiflexible resin with T300
becomes 6% smaller whereas for the rigid resin it remains constant,
when the temperature is changed from 293K to 77K.

The fracture stress of the composite in fibre direction rises
from 1700 MPa to 2010 MPa, which is an increase of 18%. Also the

fracture strain is larger at 77K than at room temperature: it rises
10% from 1.22% to 1.34%.

The mechanical properties perpendicular to the fibre measured
at 77K are low in comparison to those parallel to the fibres. The
Young's modulus is only 8%, Poisson's ratio 4%, the fracture stress
2% and the fracture strain 28% of the values parallel to the fibres.
However there seems to be a larger increase of the values due to
temperature decrease. The interlaminar shear strength τ_{ILSS} also
shows a considerable increase, when it is cooled down (Fig.13).
The fracture stress of the resin CY221/HY979 rises 70% and 90% for
the high modulus fibre and the tensile fibre, respectively.
These values are achieved at 218K, the value at 77K is somewhat
smaller. If the resin is post-cured there is a gain in the shear
strength of 73 MPa whereas the semiflexible has 49 MPa which means
an increase of 50%.

The thermal expansion of unidirectional laminates with the
semiflexible resin and the high modulus fibre M40A shows a small
value of $\overline{\alpha_t||} = 10^{-7}$ K^{-1} parallel to the fibre (Fig.14). The bar
means that it is a secant value. For 40 Vol % this value rises to
$5,3 . 10^{-7}$ K^{-1}. Perpendicular to the fibre the thermal expansion is
much higher $\overline{\alpha_t|} = 2,8. 10^{-5}$ K^{-1}. It should be noted, that the
thermal expansion coefficient $\alpha_t = \frac{d\varepsilon}{dT}$ is not constant.

INVESTIGATION ON LUGS FOR A CRYOGENIC TANK FOR
SPACE APPLICATION

For a cooled infrared laboratory (GIRL) a liquid helium tank is
needed. The tank has the form of a torus. The inner tank is suspended by 12 carriers, which have also to support the 3 radiation
shields. As the experiment needs most of the evaporative enthalpy
the losses due to thermal conductivity and thermal radiation must
be minimised. This means that the suspension elements should be
very long, have a small cross-section and a small thermal conductivity
On the other hand the suspension elements are exposed to sinusoidal
and acoustic fatigue loads during the take-off. The suspension
elements are lugs, produced with a unidirectional laminate, connected
with bolts (Fig.15). At the three upper bolts the radiation shields
are fixed. Due to friction at the bolts, interlaminar shear stresses
are induced, which cause an interlaminar shear failure. After the
crack has reached a certain magnitude, the fracture of the fibre
occurs usually at the beginning of the straight part of the lug
(Fig.16).

The fracture under static load occurs at an angle of 15° in the
round part, which is also predicted theoretically by finite element
analysis. Fatigue measurements were performed with many of these
lugs at 293K and at 4.2K (Fig.17). The stress cycle measured at

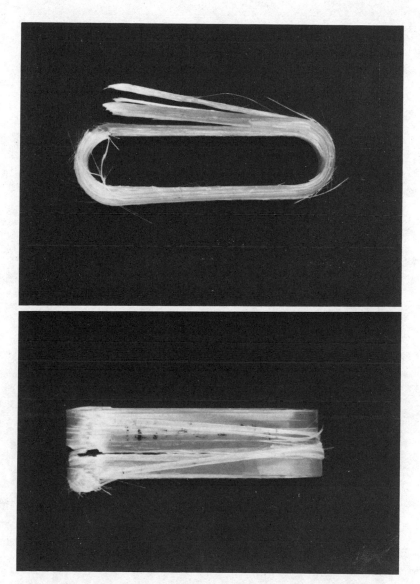

Figure 16. Rupture of GFRP-Lug due to fatigue load

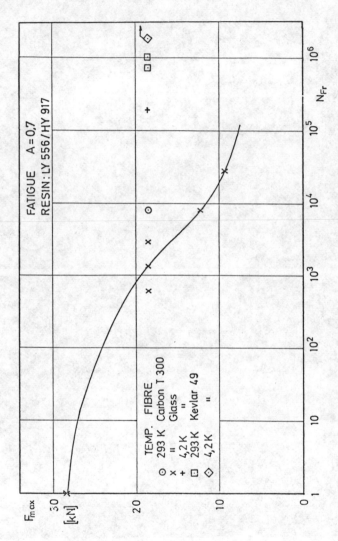

Figure 17. Stress-Cycle Diagram of GFRP and comparison of different materials at 293 K and 4.2 K

293K with glass-lugs has a good statistical base at the 19kN load.
At this point several resin systems show the same fatigue strength and
the same type of fracture. The number of cycles to failure of lugs
with aramid fibres (Kevlar) is about thousand times greater than the
one of lugs with glass fibres and hundred times greater than the one
with carbon fibres. The lugs out of glass survive at 4.2K the one
at 293K nearly hundred times. The lug out of Kevlar is a run-out-
specimen at 4.2K. Therefore the fatigue life at 4.2K is at least
twice the life at 293K.

CONCLUSIONS

    Most mechanical properties of unidirectional Carbon Fibre
Reinforced Composites show higher values at low temperatures. The
interlaminar shear strength appears to reach a maximum at 218K.
Results of fatigue tests with glass-, carbon- and aramid composites
(Lugs) show a remarkable increase in fatigue life at low temperatures
relative to room temperature.

ACKNOWLEDGEMENT

    This work is supported by the Bundesministerium fur Forschung
und Technologie (BMFT) of the Federal Republic of Germany. The
measurements were performed by the Institut fur Technische Physik of
the Kernforschungszentrum Karlsruhe.

REFERENCES

    (1)   Hartwig, G, 'Tieftemperatureiegenschaften von Epoxidharzen',
          Progr. Colloid and Polymer Science, 64,56-67 (1978), p.660

    (2)   Jones, R M, 'Mechanics of composite materials', McGraw
          Hill (1975)

    (3)   Worndle, R, 'Beitrag zur Bestimmung geeigneter Zug- und
          Druck- proben aus KFK', MBB, TN-DE133-9/78

THE USE OF GRAPHITE/EPOXY COMPOSITES IN AEROSPACE

STRUCTURES SUBJECT TO LOW TEMPERATURES

K A Philpot and R E Randolph

Hercules Incorporated
Magna, Utah, U.S.A.

INTRODUCTION

This paper will briefly describe two graphite/epoxy structures
and the critical performance parameters and properties related to
each of them. The programmes are (1) Applications Technology
Satellite (ATS) Reflector Support Truss, and (2) Side Brace for
the A37B Aircraft Main Landing Gear. The first programme was
sponsored by NASA Goddard and the second programme was sponsored by
Air Force Flight Dynamics Laboratory (AFFDL) of Wright Patterson
Air Force Base, Ohio, USA. Hercules conducted the design, development, fabrication, and testing of the structures in cooperation with
Fairchild Industries, and AFFDL.

ATS REFLECTOR SUPPORT TRUSS

The reflector support truss for NASA's Applications Technology
Satellite (see Figure 1) was required to maintain the accurate
positioning of the reflector dish with respect to the earch viewing
module (EVM). The truss met all goals including a 5 year operational
life in geosynchronous orbit. The low thermal coefficient of
expansion of graphite/epoxy minimized truss thermal growth (or
shrinkage) and provided accurate positioning of the reflector with
respect to the EVM over the expected temperature range of -160°C
(-260°F) to 93°C (200°F).

Truss weight was also of major concern, as is common in
satellite structures; and the very high unidirectional modulus/
density ratio of graphite/epoxy 114 x 10⁵m (450 x 10⁶in) versus
254 x 10⁴m (100 x 10⁶in) for aluminium offered significant weight
savings. The truss is a stability-critical structure and consists

311

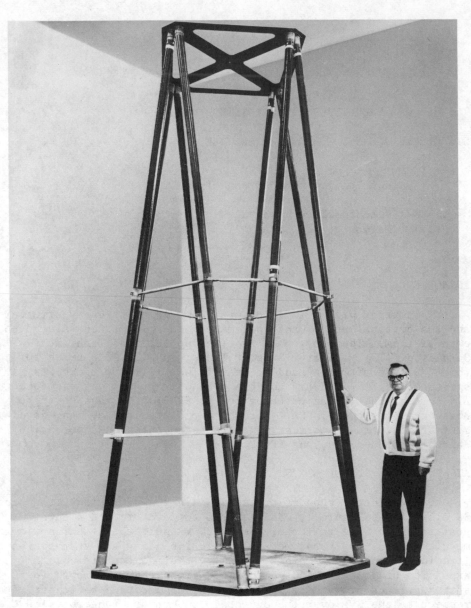

Figure 1.   Applications Technology Satellite Reflector Support Truss

of eight graphite epoxy tubes 4.42 m long by 6.35 cm ID with 0.254 cm ID (14.5 ft long by 2.5 in. ID with an 0.1 in wall thickness).

The truss tubes were each a hybrid construction using 2002/HMS graphite and 2002/S-901 glass. The fabrication process involved a combination rolling table lamination and filament winding technique. The tubes were bonded to each other ghrough titanium end fittings using room temperature cure EA-934 adhesive.{1}

The truss was the first graphite epoxy primary structure to be used in a NASA satellite. The approaches for solving the design and fabrication problems were, therefore, conservative. Even with conservatism, the graphite/epoxy prototype truss, which was success- fully tested to static qualification levels, weighed 36.7 Kg (80.8 pounds) compared to an aluminium truss weighing 77.1 kg (170 pounds). Development testing at all environmental conditions (space and earth) indicated full compliance with all truss requirements. Dynamic testing on the full-scale prototype spacecraft was successfully completed by Fairchild Industries.{2}

Engineering data on 2002/HMS material upon which the truss was designed are presented below.

Tensile and Compression Tests

Standard tensile tests were performed at -162°C (-260°F), 25°C (77°F) and 93°C (200°F) using end-tabbed 0° longitudinal specimens. Compression tests were conducted using 0° sandwich beams. From Figure 2 it is seen that the modulus changed very little with temperature and tensile strength was maximum at room temperature. Transverse tensile and compression data are shown in Figure 3 with tensile strength changing only slightly with temperature, while compression strength was maximum at -162°C (-260°F).

In-Plane and Interlaminar Shear Tests

In-plane and interlaminar shear tests were also conducted at -162°C (-260°F), 25°C (77°F) and 93°C (200°F). The data are shown in Figure 4 and shows that in-plane shear modulus decreases as temperature increases while interlaminar shear strength remained constant at temperatures less than 25°C (77°F) but dropped off at higher temperatures.

Flexural Strength and Modulus Tests

Flexural tests performed at -162°C (-260°F), 25°C (77°F) and 93°C (200°F) showed relatively small changes with temperature as shown in Figure 5.

TABLE 1.   THERMAL CONDUCTIVITY TEST RESULTS

Mean Temperature		Thermal Conductivity $wm^{-1} K^{-1}$	
K	°C	90° Transverse	0° Longitudinal
203	-70	0.69	31.5
253	-20	0.78	35.5
273	0	0.82	37.0
293	+20	0.85	38.4
333	+60	0.89	41.5
373	+100	0.93	44.4
423	+150	0.95	47.5

Note: Specimen size -9 mm x 63 mm x 63 mm, average specimen
      density 1559 kg m^{-3}

TABLE 2.   EFFECT OF THERMAL CYCLING

Tensile Property	Before Cycling	After Cycling
0° tensile, (Mn/m^2) ksi	914.3 (132.7)	780 (113.2)
0° strain, %	0.51	0.49
90° tensile, (Mn/m^2) ksi	7.15 (49.2)	7.26 (50.0)
90° strain, %	0.80	-

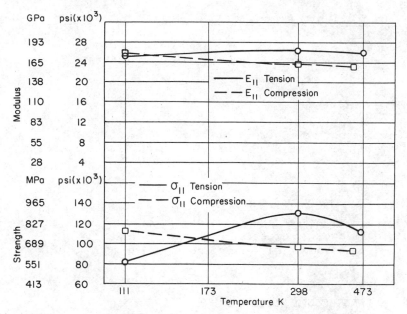

Figure 2. ATS2002 HMS Unidirectional Properties Versus Temperature

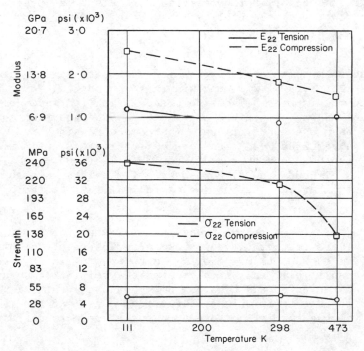

Figure 3. ATS2002 HMS Transverse Properties vs Temperature

Thermal Coefficient of Expansion and Thermal Conductivity
──

Thermal coefficient of expansion data and thermal conductivity
data are shown in Figure 6 and Table 1 respectively.

Thermal Cycling
───────────────

Tensile specimens were cycled 365 times between -73°C (-100°F)
and 93°C (200°F) at 10^{-6} torr with the results shown in Table 2.

Double Lap Shear Testing
────────────────────────

Table 3 summarizes the results of adhesive shear testing and
environmental cycling effects. It appears that the particular
exposure did not adversely affect bond strength.

A37B MAIN LANDING GEAR SIDE BRACE

The purpose of the A37B main landing gear side brace programme
was to demonstrate the design, fabrication, and performance
capabilities necessary for composite materials to be used in cyclic,
highly concentrated load applications, specifically, aircraft landing
gear parts. Figure 7 shows a typical A37B graphite/epoxy side brace
assembly.

Hercules manufactured 50 graphite composite landing gear side
brace assemblies using filament winding, lamination, warm forming,
and closed mold curing. The side brace assemblies were composed
entirely of AS graphite fibre and modified 3501 epoxy except for
aluminium bushings and fittings. The composite side brace was shown
to have the required load transfer characterisitcs and ability to
withstand the high concentrated loads(30.7 kN (6900 pounds)tension
and 66.3 kN (14,900 pounds) compression). The design also withstood
six fatigue lifetimes and was successfully flight tested by the
USAF.{3}

The upper side brace was subjected to extensive environmental
testing. These tests included exposure to 95% relative humidity at
49°C (120°F) to 93°C (200°F) for 9 months;soaking in JP4, anti-icing
fluid, hydraulic oil, and trichloroethylene solvent; exposure to
ultraviolet radiation and high humidity; natural weathering; abuse
with hand tools; six lifetimes of fatigue loads; and the "Rangoon
cycle" involving cyclic exposure to temperature extremes and high
humidity.

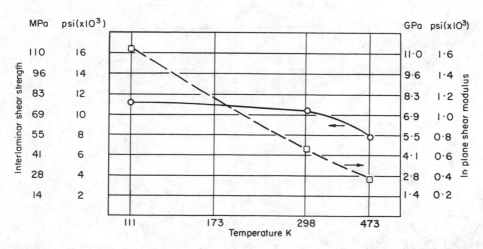

Figure 4.   ATS2002 HMS Shear Strength and Modulus Versus Temperature

TABLE 3.   ENVIRONMENTAL EXPOSURE EA-934 DOUBLE LAP SHEAR RESULTS
           25°C (77°F)

Not Cycled		Cycle, -73°C to 93°C (-100°F to 200°F)	
Sample No.	Bond Failure Stress (Mn/m²) psi	Sample No.	Bond Failure Stress (Mn/m²) psi
		E-34	(31.0) 4500
DLS-6	(23.4) 3400	E-35	(23.4) 3400
DLS-8	(16.2) 2350	E-36	(20.4) 2960
DLS-10	(22.2) 3225	E-38	(27.8) 4040

*Cycles 365 times at 10⁻⁶torr

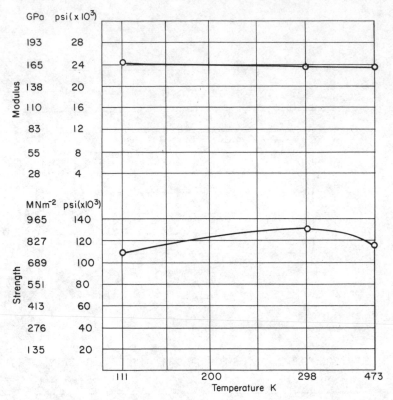

Figure 5.   ATS2002M Undirectional Flexural Properties Versus Temp-
            erature.

Baseline Testing

Baseline testing of dry specimens was run to establish a
reference point for the environmental tests. As shown in Figure 8,
the compressive strength was not degraded at the lower test
temperature; increased strength values were observed.

Moisture Absorption

Moisture absorption effects are shown in Figure 8. The compressive failure strength was unaffected by the lower test temp-
earture. Figure 11 also gives some moisture absorption data. At
room temperature 25°C (78°F) and -54°C (-65°F) compressive failure
load was unchanged with added moisture up to 1.3%.

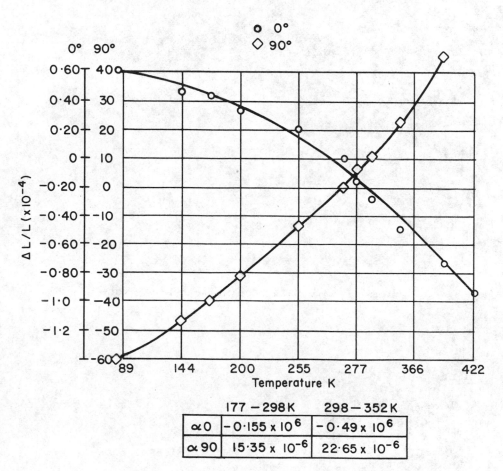

Figure 6. 2002/HMS Graphite Epoxy Thermal Expansion Data.

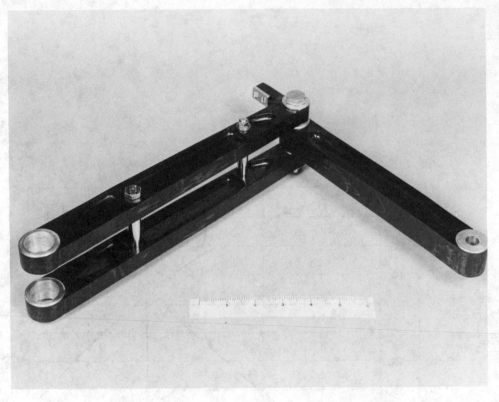

Figure 7.   A37B Main Landing Gear Side Brace

Cycling

Test specimens undergoing the "Rangoon Cycle" were exposed to:
(a) 95% RH, 49°C (120°F), 21 hours
(b) Room conditions, 20 minutes
(c) -54°C (-65°F), 1 hour   1 cycle
(d) Room conditions, 20 minutes
(e) 93°C (200°F), 1 hour
(f) Room conditions, 20 minutes

The cycling was extended over 62 calendar days and included 21 full cycels. Excess time was spent in the humidity chamber.

As shown in Figure 8, the strength of the cycled items increased at -54°C (-65°F). The samples tested at 74°C (165°F) showed an average strength decrease of 27%, but strength remained above the design ultimate load value.

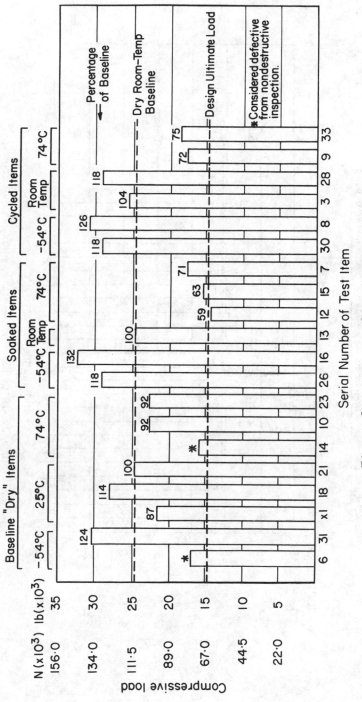

Figure 8. Summary of Failure Loads

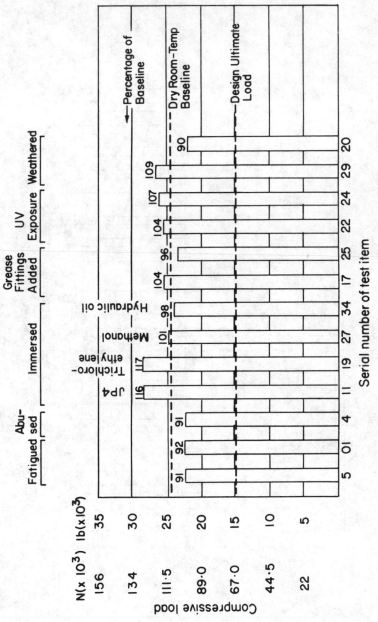

Figure 9. Summary of Loads and Environmental Exposures

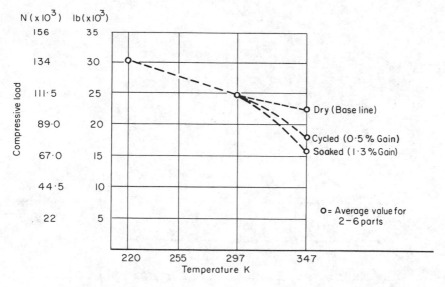

Figure 10.   Failure Loads Versus Temperatures

Fatigue

The two fatigued upper side braces had been subjected to four aircraft lifetimes (100,000 cycles) plus two additional lifetimes at 25% overload tested in compression at room temperature, they still retained 92% of the ultimate strength.

Abuse

One side brace was subjected to severe intentional abuse deisgned to simulate abuse by maintenance personnel and foreign object damage.

It was observed that the part seemed almost indestructible at first. It sustained very little visible damage considering the abuse, and appreantly the ultimate strength did not suffer appreciably (91% baseline strength). Failure occurred at the small bushing.

Immersion Test

Four side braces were totally immersed in various fluids, but absorption was negligible in all cases. The following schedule was

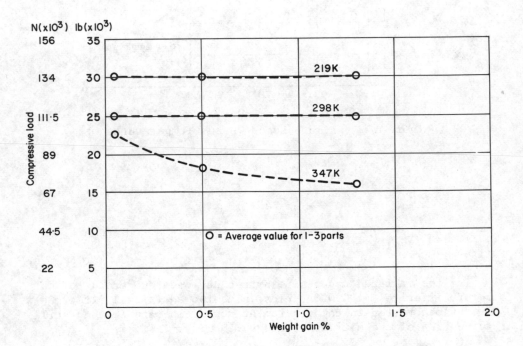

Figure 11.   Failure Loads Versus Moisture Gain

observed:
(a) JP4 immersion, 17 days (jet fuel)
(b) Trichloroethylene, 24 hours (common degreaser solvent)
(c) Methanol, 17 days (anti-icing fluid)
(d) Hydraulic oil, 17 days

Ultimate strengths of these parts after immersion ranged from
98% to 117% of the dry, room temperature baseline value as shown in
Figure 9.

Weathered

The weathering condition involved exposure of side braces to
the Ohio climate from September 1973 to April 1974. Figure 9 shows
the test results of the weathered specimens.

Figure 10 summarizes the temperature effect on the compressive
strength of the upper side brace for the dry, cycled, and soaked test
specimens. Low temperature testing $-54^{\circ}C$ $(-65^{\circ}F)$ resulted in a
compressive strength increase. Strength degradation was observed
at $74^{\circ}C$ $(165^{\circ}F)$.

Failure load versus moisture gain (Figure 11) data indicates
that moisture absorption below 1.3% has no effect on compressive
strength at room temperature and $-54^{\circ}C$ $(-65^{\circ}F)$ testing temperatures.

CONCLUSIONS

Graphite composite structures have been shown to be reliable
over wide temperature ranges. Not only do graphite composites
result in large weight savings over metal counterparts, but they are
also quite comparable in strength properties. Experimental research
as well as actual applicaiton useage have shown that graphite
composites can be used successfully for high performance applications.

REFERENCES

(1) Randolph, R.E., Jensen, L. C., Martin, W.,"Graphite/Epoxy
    Reflector Support Truss for Applications Technology Satellite
    (ATS)", Hercules Incorporated, Bacchus Works, Magna, Utah,
    October 1972

(2) McNeill, C. E., "Composite Material Characterization for Large
    Space Structures", Fairchild Space and Electronics Division,
    Germantown, Maryland, August 1972

(3) Dexter, Peter F., Lt USAF, "Graphite/Epoxy Landing Gear
    Environmental Tests", Technical Memorandum AFFDL-TM-217-FEM,
    Air Force Flight Dynamics Laboratory, Wright Patterson AFB,
    Ohio, November 1974.

STANDARDIZING NONMETALLIC COMPOSITE MATERIALS

FOR CRYOGENIC APPLICATIONS

M B Kasen

Fracture and Deformation Division, National Bureau of
Standards,
Boulder, Colorado 80302, U.S.A.

INTRODUCTION

The technology of composite materials is presently at the state
that metals technology was before being systematized by the establishment of industry standards. It is difficult to imagine working
in metals technology in the United States without the ability to
associate performance with designations such as 304L for stainless
steel or 2024-T4 for aluminium. It would be equally difficult for
individuals in other countries to perform effectively without
similar national metal and alloy coding systems that transcend
proprietary designations, but this is the situation presently con-
fronting those working in the composite materials area, in the United
States and elsewhere.

Up to the present time, the problem has not been of major
concern because the consumption of high-performance composites has
been relatively low in the consumer area. However, consumption is
expected to increase greatly as escalating energy costs place pre-
miums on light-weight fuel-efficient transportation systems and the
development of alternative energy sources. It will become increasingly difficult to meet the future needs of composite technology
with proprietary systems or with laminates fabricated from the few
formulations published in the literature.

The need for composite materials standards is therefore common
to many industries. As reflected in the forum chaired by Evans
at the 1979 ICMC/CEC conference[1], the need of the cryogenic industry
is particularly acute. This industry is already at the point where
many of the nonmetallic materials needs of magnetic fusion energy
(MFE), magnetohydrodynamic (MHD), and rotating cryogenic machinery

327

systems cannot be adequatly met by existing commercial products.

A primary problem is the inability to associate reliably the
cryogenic performance required for diverse applications with the
profusion of products identified by trade names, most of which were
developed for other purposes. This paper outlines the current
thinking of the author on this overall problem and describes the
efforts currently underway to meet the immediate industrial cryogenic
requirements. The author hopes that these considerations may be
combined with similar considerations based on European experience
to contribute to the eventual establishment of workable international
standards for nonmetallic materials.

The author will concentrate on nonmetallic composite materials,
reflecting the current need for such materials in the fabrication
of the cryogenic portions of new energy sytems. It is recognized,
however, that unreinforced polymers pose a similar standardization
problem.

IMMEDIATE CONCERNS

The immediate task is to assist designers and fabricators in
the selection of dependable insulating and structural nonmetallic
laminates for MFE and MHD systems presently being built in the
United States. The high costs of these systems and the substantial
performance demands made on the laminates require that the selected
materials be well characterized at cryogenic temperatures and the
variability in performance be minimized. Because the present systems
will provide experience that can be extrapolated to more sophisti-
cated, larger systems, the selected materials must be commercially
available over a period of years. Furthermore, they must lend themselves to a systematic materials development programme in response
to industry needs. These boundary conditions imply that control
over the constitutive elements, manufacturing processes, and performance characterization will be necessary.

Current construction will necessarily draw on state of the
art, commercially available nonmetallic laminate materials developed
for other than cryogenic service. In the past, many proprietary
products have been used. However, this approach is inefficient for
demanding applications since it requires that each group responsible
for a design conduct an expensive in-house materials characterization
programme. It must also be assumed that the selected proprietary
product will not be substantially altered in response to market
forces outside the cryogenic area. The alternative of purchasing
laminates to specified cryogenic performance criteria either requires
the manufacturers to develop cryogenic quality control procedures
or assumes that room temperature data can be confidently extrapolated
into the cryogenic range. Unfortunately, the latter concept is
rudimentary for most conventional properties and nonexistent for

special performance requirements under cryogenic irradiation.

A third alternative is to establish uniform industry specifications for manufacturing a series of well-characterized nonmetallic
laminates having the greatest application to cryogenic technology.
This approach recognizes that many of the inexpensive commercial
laminates developed for room temperature applications are useful
cryogenic materials, whose primary dificiencies are the lack of a
cryogenic data base and an excessive variability when used in a
temperature region for which they were not intended. Establishment
and characterization of uniform specification laminates addresses
these deficiencies and provides designers with materials having
greatly improved cryogenic reliability at minium cost. Furthermore,
the uniform specifications assure that the critical material factors
affecting performance will be controlled by the cryogenic industry,
providing the basis for a systematic materials development programme
that reflects the needs of the industry.

We have been following this approach in working with the
United States laminating industry to develop a series of standard,
high-pressure industrial laminates for cryogenic service. Both the
industry and the coordinating standards group, the National Electrical Manufacturers Association (NEMA) have been most cooperative.
Thus far, two grades of glass-fabric reinforced epoxy laminates have
been established, characterized, and placed in commercial production.
These materials are designated G-10CR and G-11CR to distinguish them
from the conventinnal NEMA G-10 and G-11 products. Performance data
for these cryogenic grades have been published in the literature[2].
Acceptance by the cryogenic industry in the United States has been
gratifying. NBS is continuing to work with industry to develop
additional standard specifications for other laminate systems of
interest to cryogenics, possibly including random-mat and uniaxial
glass-reinforced epoxy matrix laminates.

These developments are meeting the immediate needs of industry
while providing the basis for systematic alternation in the performance of standard materials in response to developing industrial
needs. The concept of the overall programme is outlined by the
flow chart of Figure 1. Built-in iteration steps provide the open-
ended type of programme required to be responsive to continually
developing needs. The sequence depicted for development of the
CR grades of G-10 and G-11 illustrates the progress up to the
present time. The characterization data indicate that these products
will perofrm very well in most applications.

Studies at the Oak Ridge National Laboratory, however, have
shown that the mechanical properties of these laminates are severely
degraded by 4 K neutron and gamma irradiation at fluences comparable
to those expected in a functioning MFE reactor[3,4]. This deficiency
is being addressed by a subprogramme to develop laminates having

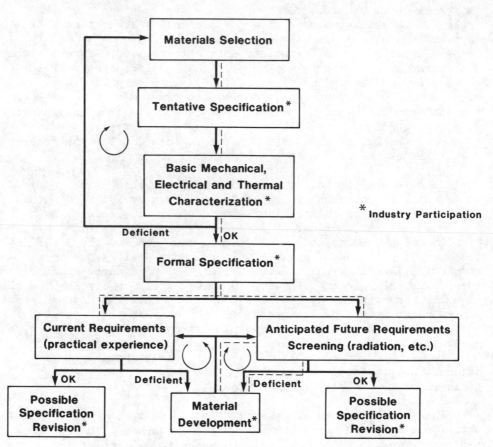

Figure 1.   Flow chart depicting a proposed approach to meeting
immediate needs of nonmetallic laminate development
for the MFE, MHD and rotating cryogenic machinery
industries. Dashed lines depict the present state
of development of the G-10CR and G-11CR grades of
industrial laminates.

improved radiation resistance. The first step will be to determine
if radiation resistance will be significantly improved by replacing
the epoxy matrix of the G-10CR laminate with a polyimide matrix.

It appears to the author that such a programme would also
ential an iterative cycle like the one illustrated in Figure 2.
Because the cryogenic radiation resistance of a polymer is likely
to be a strong function of the exact molecular structure[5], an
essential part of the cycle is a correlation of molecular level
damage with radiation level and property change. Only in this way
will the required scientific basis be established. The programme
envisions the development of methods for preliminary screening of
4 K radiation resistance in the laboratory prior to confirmatory
testing in a reactor. It is evident that considerable new ground
will have to be broken in solving the radiation problems in non-
metallic materials.

A major deficiency in the approach thus far discussed is the
absence of an industry standards group having the responsibility
for coordinating, updating, and disseminating information on
developed standards. The NEMA organization is performing the
disseminating function for the CR-grade laminates through their
Industrial Laminates Subcommittee. But NEMA interest is confined
to high-oressure, bulk laminates. An organization responsive to
the broader interests of the composite materials industry is required.

LONG-RANGE STANDARDIZATION NEEDS

The above approach lacks the comprehensiveness required to
assure orderly overall composite materials development and
implementation in the long run. The author agrees with Evans that
the most effective way of alleviating the present chaotic state of
nonmetallic materials technology would be to introduce a numerical
calssification system based upon key elements of a material that
affect performance[1]. Here, the cryogenic industry has much in
common with other industries exploiting composite technology.

There seems no a priori reason why such an approach would not
be feasible for composite laminates, when one recognizes that the
significant engineering properties of a laminate at any temperture,
are largely defined by relatively few parameters. For example,
the intrinsic mechanical, elastic, thermal and electrical properties
of a laminate should be defined within relatively narrow limits by
a numerical classification system that establishes the type and
flexibility of the matrix, the type and configuration of the
reinforcement, and the reinforcement volume fraction. At the very
least, it should be possible to establish meaningful lower bounds
on such properties. Special properties, such as radiation resis-
tances, could continue to be associated with controlled-specification
products that could be incorporated into the basic coding system.

TABLE 1.   LAMINATE CLASSIFICATION OF PRIMARY
INTEREST TO CRYOGENIC TECHNOLOGY.[+]

Laminate Category	Suggested Coding
Uniaxial Glass-Epoxy	E1U55U-E*
	E1U55U-S2
	E1U55U-S
Glass Fabric-Epoxy	E1F48U-E/76*
	E1F40U-E/76
	E1F40U-E/XX
Glass Mat-Epoxy	E1M35U-E*
Glass Fabric-Polyester	PE1F45U-E/76
	PE1F40U-E/76
	PE1F40U-E/XX
Glass Mat-Polyester	PE1M35U-E
Glass-Polyester Pultrusion	PE1F4U-E/11(P)*
Cotton Fabric-Phenolic	PH5XXU
Glass Fabric-Melamine	M1F45U-E/76
	M1F40U-E/XX
Uniaxial Aramid-Epoxy	E2U60U-49
Aramid Fabric-Epoxy	E2F50U-49/94
Uniaxial,High-Strength Graphite-Epoxy	E3U60U-HS
Fabric, High-Strength Graphite-Epoxy	E3F60U-HS/100
Uniaxial, Medium-Modulus Graphite-Epoxy	E3U60U-M
Uniaxial, High Modulus Graphite-Epoxy	E3U60U-HM
Uniaxial, 5.6-mil Boron-Epoxy	E4U50U-56

[+]For illustration purposes only

*Classifications meriting development of special cryogenic grades

It is beyond the scope of this paper to suggest a definitive system, but it is worthwhile to consider the general form that such a coding or classification system might take. One possibility, paralleling metals practice, is illustrated in Figure 3. Here, the first letter(s) and number define the matrix and reinforcement class, identifying the composite as being in the general category of epoxy-glass, polyimide-graphite, etc. A following letter defines the reinforcement configuration, e.g., E1U would define a uniaxially reinforced,epoxy-glass composite. The next two digits define the nominal fibre volume fraction. A ±3% range would be sufficiently precise and would be compatible with existing manufacturing practice. An E1U60 coding thus defines a 60 ±3% volume fraction uniaxial glass-epoxy laminate. The performance of such a laminate could be quite definitively established by adding general information on the degree of matrix flexibility and specific information on the type of reinforcement. Thus, a designation E1U60U-E would not only define a 60% volume fraction uniaxial E-glass reinforced, fully reacted epoxy matrix laminate but would simultaneously establish meaningful intrinsic performance bounds on the laminate performance. Extrinsic factors related to laminate quality, such as void content and interply bond integrity, will affect performance, but such factors must be independently assessed for all laminates.

The fabric used to reinforce laminates is frequently un-balanced, usually having a lower fibre content in the fill direction of the weave than in the warp direction. For example, a type of 7628 glass weave commonly used in the United States has a fill/warp ratio of 32/42. In the scheme of Figure 3, information on this unbalance is provided by a two-digit number defining the fill/warp fraction, or 76 in the cited case.

This type of system offers flexibility in establishing coding at the specific level required for the application of interest. The cryogenic industry would probably be reasonably well served by the categories listed in Table 1. The list could be expanded or individual designations could be made more or less specific as need arises. Some cryogenic data already exist for most of the materials on this list, although in no case are the data sufficient for establishment of statistically significant lower performance bounds.

A standardization scheme of this type would simplify materials selection and purchase; furthermore, it would minimize costs by providing the maximum flexibility to the laminating industry in meeting the requirements.

Consider, for example, the purchase of a G-10 type of laminate widely used in superconducting magnet construction in the United States. Under the present system, the cryogenic performance of this type of laminate purchased on the open market can vary by 30%, if produced by a reliable laminator, or more, if produced by

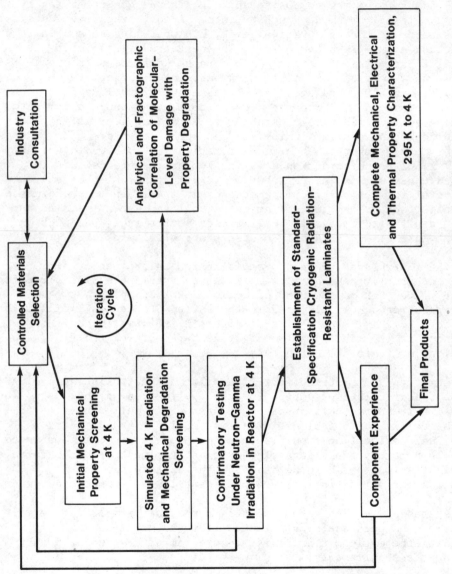

Figure 2.   Suggested flow chart for development of radiation–
resistant nonmetallic laminate insulators.

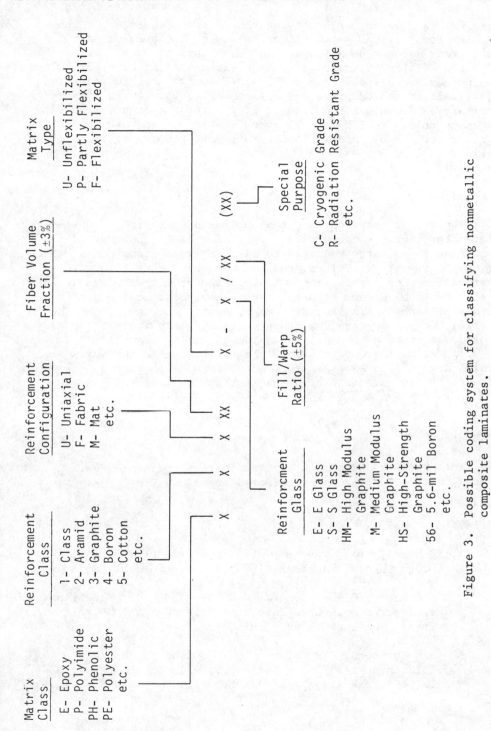

Figure 3. Possible coding system for classifying nonmetallic composite laminates.

marginal fabricators. Specifying the CR grade would solve the
problem for critical applications; however, the premium paid for
this grade may not be justified for many applications. Here, the
illustrated classification system provides a solution by allowing
the purchaser to specify an E1F48U-E/76 laminate if the generic
equivalent to the CR grade was desired or to specify an E1F40U-E/76
laminate as a minimum requirement if greater property variability
is acceptable. An E1F40U-E/XX specification would result in still
further economy by removing the requirement for a specific fill/warp
ratio.

SUMMARY

The author has described the ongoing efforts in the United
States to provide a commercial supply of specific controlled-
performance laminates required to meet the short-term needs of the
cryogenic industry. This approach also provides the basis for a
systematic materials development programme required to meet future
needs, but the current effort addresses only the most demanding
useage of nonmetallic laminates in cryogenic construction. The
overall long-range requirements of the cryogenic industry justify
consideration of a more comprehensive system of nonmetallic laminate
classification that would simplify materials selection while
minimizing costs. An example of such a system is provided to
illustrate its potential benefits to both the consumer and the
producer.

ACKNOWLEDGMENT

This work was supported by the Office of Magnetic Fusion
Energy, United States Department of Energy.

REFERENCES

1. D Evans, in "Advances of Cryogenic Engineering," Vol. 26,
    Plenum Press, New York (1980), p. 83 (in publication).

2. M B Kasen, G R MacDonald, D H Beekman, and R E Schramm
    in "Advances in Cryogenic Engineering", Vol. 26, Plenum
    Press, New York (1980) (in publication).

3. R H Kernohan, R R Coltman and C J Long, "Radiation
    Effects on Insulators for Superconducting Magnets", ORNL/
    TM-6193, Oak Ridge National Laboratory, Oak Ridge, TN (1978)

4. R H Kernohan, R R Coltman and C J Long, "Radiation Effects on Organic Insulators for Superconducting Magnets", ORNL/TM-7077, Oak Ridge National Laboratory, Oak Ridge, TN (1979).

5. B S Brown, "Radiation Effects in Superconducting Fusion Magnet Materials", submitted to J. Nucl. Mat. (1980).

SPECIAL APPLICATIONS OF NONMETALLIC MATERIALS

IN LOW TEMPERATURE TECHNOLOGY

W Elsel

Siemens Research Laboratories
Erlangen
Federal Republic of Germany

INTRODUCTION

In large scale superconducting technical devices, it is often
necessary to transmit large forces from low temperature to room
temperature. The thermal losses caused by the cryogenic structure
essentially determine, in some cases, the efficiency of the total
superconducting system. Therefore these losses have to be minimised.
This can be done in two ways:

(1)   constructively: with distances between the temperature
       levels as large as possible and a thermal conducting cross-
       section compatible with operating stresses; for compressive
       loads, buckling has to be avoided without enlarging the
       thermal cross-section;

(2)   using structural materials with a low ratio of thermal
       conductivity to strength.

Examples of the successful application of fibreglass reinforced
epoxide resins as strut materials for a stellarator prototype magnet
and for levitation magnets will be described.

From design studies that are in progress for large superconducting inductive energy systems, it is shown that nonmetallic
materials will be increasingly important in low temperature techniques in the future.

MATERIAL TESTS

Suitable support materials were chosen by testing self

339

Table I.   Mechanical and thermal properties of selected materials

properties / material	temperature K	Epoxy/ glassrov. 70% wt. glass ① II roving	Epoxy/glass cloth 55% wt.glass ② II cloth	⊥ cloth	Phenolic- resin papertyp Hp 2061 ⊥ sheet	Steel X2Cr NiMo 18/14 0.2 N	Tita- nium alloy 5Al,2.5Sn ③
		σ_c	σ_c	σ_c	σ_c	$\sigma_{t\,0,2}$	σ_t
stress σ MPa	300	490	470	500	275	350	840
compressive σ_C	78	750	760	800	660	750	1450
tensile σ_t	4.2	970	790	900	–	1000	1820
strain ϵ %	300	1.7	2.7	6.3	7.2	0.36	1.0
according the	78	2.4	4.0	6.3	4.3	0.55	1.5
above stresses	4.2	–	–	–	–	0.67	2.0
thermal dilatation	300/78	1.33	2.66	6.0	3.6	2.86	1.65
$\Delta L/L$ ‰	78/4.2	0.26	0.81	0.93	0.53	0.24	0.15
integral therm. conductivity $\int_{T_1}^{T_2}\lambda\,dT$ W/cm	300/78	0.96	1.26	0.77	1.7	27.1	22.8
	78/4.2	0.15	0.19	0.13	0.24	3.5	3.7
$\dfrac{1.5\int_{T_1}^{T_2}\lambda\,dT}{\sigma(T_2)}$ $\dfrac{\mu Wcm}{N}$	300/78	29.4	40.2	23.1	92.7	1161	407
	78/4.2	3.0	3.75	2.4	5.5	70	38.2

① resin: Epikote 162/Laromin C 260, 100 : 85, with wrapped
   roving, glass content for the inner cylinder
② HgW 2374.2, EGS 104 (Ferrozell)
③ Cryogenic Materials Data Handbook Volume II, AD 713619, 1970

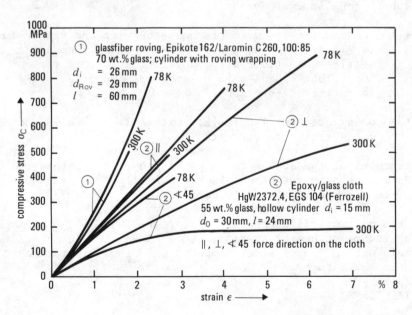

Figure 1.   Stress-strain behaviour of fiberglass epoxies

fabricated as well as commercially available composites. Mechanical and thermal properties have been measured at 4.2 K, 78 K and 300 K on cylindrical specimens of fibreglass epoxide and phenolic resin with paper. Results are given in table 1.[1] For comparison, data on a stainless steel and a titanium alloy are also presented.

For given distances between fixed temperature levels T_1 and T_2 and a fixed load the heat flow through the supporting material is determined by $\int_{T_1}^{T_2} \lambda \, dT/\sigma$; where λ is the thermal conductivity and σ the compressive or tensile stress. For the composites, a safety factor of 1.5 was introduced with regard to the short-time ultimate compressive strength, and for metals the 0.2% proof stress was taken. The heat flow through composites is some 10 to 50 times less than through the various metals considered.

The compressive stress – strain behaviour of fibreglass epoxies at 78 K and 300 K is shown in Fig 1. The greatest compressive modulus (and hence the smallest strain) was achieved with the unidirectional roving reinforcement (specimen 1) with the force applied parallel to the fibre direction. The rovings consist of bundles of 12 to 120 parallel glass fibres with diameters of 9 to 11 μm. Unlike the cloth-specimens (2), the roving specimens were prone to premature breaks because of local delamination. This problem can be overcome by wrapping the perimeter of the specimen with fibreglass.

The measurements showed that sudden increased load does not damage the material so long as the ultimate strength is not exceeded. With the load parallel to reinforcement, fibreglass epoxy specimens 1 and 2 were stressed at 78 K and 300 K, increasing the load from 50 to 300 MPa within 0.3 s. With the load perpendicular to cloth, the specimen 2 was stressed at 4.2 K and 78 K in 0.3 s from 100 MPa to 700 MPa. After 20 load cycles no visible damage or permanent deformation was noticed.

APPLICATIONS IN THE PAST

Stellarator Prototype Magnet

This superconducting magnet was developed by IPP Garching and Siemens as a prototype for the torus magnet of the stellarator project W7. The technical concept of the prototype module has been described elsewhere.[2]

The basic layout of the magnet is presented in Fig 2. Axial forces are transmitted from the winding via the helium enclosure and eight thermally insulating supports to each of the two sides facing the module casing. Hydraulic systems at the support points on the casing side provide the necessary prestress. The axial supports are capable of withstanding the maximum force of 2.5 x

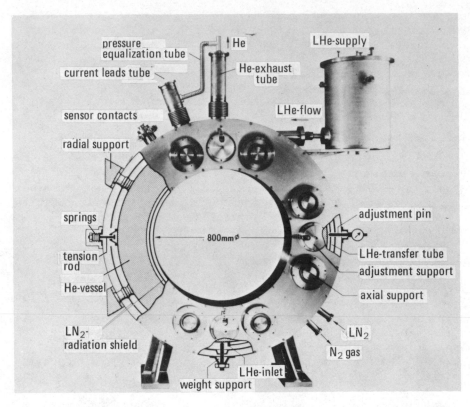

Figure 2. W7 prototype-magnet cryostat

10^6 N exerted per coil in the torus assembly, if the magnetic
equilibrium is disturbed. The force acting on the coil in the
direction of the toroid centre amounts to 7 x 10^5 N. Two radial
supports transmit 4 x 10^5 N each from the winding to the module
casing.

    The radial and the outer axial supports consist of the material
of specimen 1 and the inner axial supports of the material of
specimen 2 with the load perpendicular to the cloth. The construction and the dimensions of an axial support are shown in Fig 3.
The other supports differ only slightly in the length. With
bellows the axial supports can be reset if the hydraulic is relieved
of the load. Then they are thermally uncoupled. The supports are
fixed at the warm end, whereas there are slide faces at the cold
end. The safety factor against fracture amounts to 1.6 for the
axial supports and 2.5 for the radial supports. The thermal losses
of the total support system are 7 W at 4.2 K and 75 W at 78 K.

The supports performed very well in a simulation experiment under axial force loads of up to 120% of the design values.

Levitation Magnets

Within the German magnetic levitation project Siemens has developed and fabricated eight superconducting magnets for the 17-ton test carrier at Erlangen. The project and its components have been described in detail elsewhere.[3]

The magnets have an identical layout, four as lift magnets and four as guiding magnets. The winding has a length of 1.0 m and a width of 0.3 m. With respect to the electrodynamic principle the winding was effected in all three axes of the space by forces which have to be transmitted to the vacuum vessel and further on to the vehicle. The supporting system is shown in Fig 4. Compressive loads are transmitted by fibreglass epoxy struts, and tensile loads by stainless steel tapes.

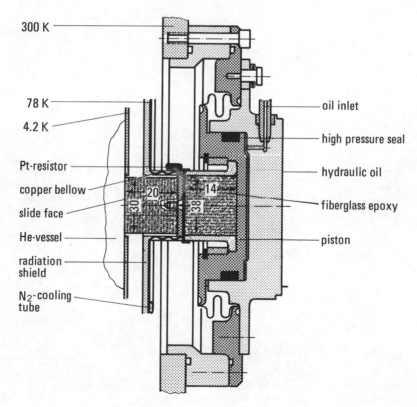

Figure 3. Support between 4.2K and 300K with hydraulic loading

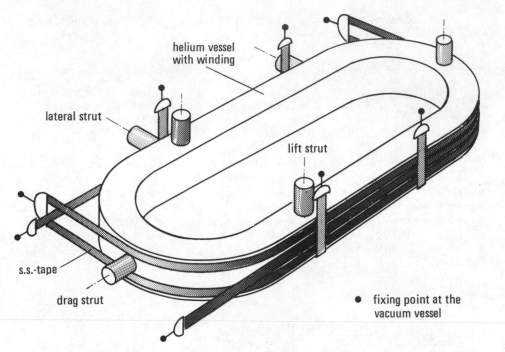

Figure 4.   Levitation magnet: mounting support of the winding

We will discuss as an example the lift struts.   The requirements
of these are:

  - to carry the vehicle mass increased by a shock factor with
    respect to dynamic loads,

  - to minimise the heat flow from the warm vacuum vessel to the
    cold He-vessel with the winding,

  - a small overall height.

The cross-section of the Heim-type [4] struts is shown in Fig 5.
The nonmetallic tubes consist of glass cloth/epoxy with 60 percent
in weight of glass.   The working load of the struts is composed of
electromagnetic, electrodynamic and inertia forces.   In addition
wind forces and prestresses are to be considered.   In the driving

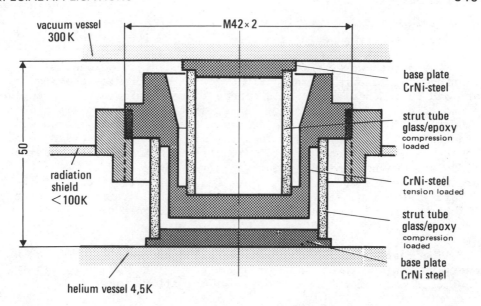

Figure 5.   Lift strut

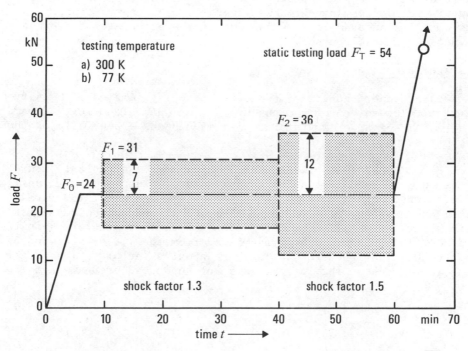

Figure 6   Testing program for the lift struts of the test carrier
            at Erlangen

mode a dynamic force, which is taken into account by a shock factor
of 1.5, is superimposed on the calculated quasistatic load.

The struts were tested under static and dynamic load. The
testing programme is shown in Fig 6. Within 5 minutes the static
preload is reached and is held constant over 5 minutes. Thereupon
a sinusoidal dynamic load of 1 Hz is superimposed. Its amplitude
amounts to 30% of the preload (shock factor 1.3) for a time of 30
minutes (1800 load cycles). Then the amplitude is increased to 50%
of the preload (shock factor 1.5). After 20 minutes the dynamic
load is withdrawn and the strut is increasingly loaded until it
breaks. Results of the testing of lift struts are given in table 2.
Before assembling, each strut was loaded with a static load which was
50% higher than the peak force in the dynamic case with a shock
factor of 1.5.

The test carrier covered an overall distance of 10,200 km, with
energised magnets over 1,000 km and a cumulated stretch of way in
levitated mode of 200 km. The struts met all requirements.

FUTURE APPLICATIONS

Nonmetallic structural material could be increasingly important
in future large scale superconducting technical devices such as
magnets for fusion, synchronous generators and energy storage
systems. As an example some aspects of the structure of storage
magnets will be discussed.

Storage Magnets

Large superconducting inductive energy systems could be very
attractive as an economic means of meeting peak electricity demands.[5]
The economy of an energy storage magnet depends essentially on the
support structure.

For a 1 GW/4 GWh-storage unit, which may be a typical future
size, we performed a study on the structure. Such a magnet consists
of one inner coil and two outer coils with a radius of 139 m resp.
165 m and a height of 40 m resp. 35 m. The inner and outer coils
were separated axially 26 m from each other. Lorentz-forces in the
order of 10^{11} N have to be transmitted from the 1.8 K-cold winding
to the 300 K-warm bed-rock. They are about a factor of 10^3 larger
than the forces of the largest existing superconducting magnets.

Table II.   Stress and strain of lift struts (glass/epoxy)

		F_0 24 kN static load	F_1 31 kN	F_2 36 kN	F_T 54 kN testing load
compressive stress σ_C MPa	inner tube	89	115	133	193
	outer tube	92	119	138	200
compressive strain of the total strut s mm	300 K	1.5 ± 20%	0.29 ± 15%	0.55 ± 15%	–
	78 K	0.76 ± 30%	0.26 ± 20%	0.46 ± 20%	–

Fracture behaviour at 78 K of a strut sample

force F_B		90.5	kN
stress σ_{CB}	inner tube	335	MPa
	outer tube	348	MPa
strain of the strut s		4.85	mm

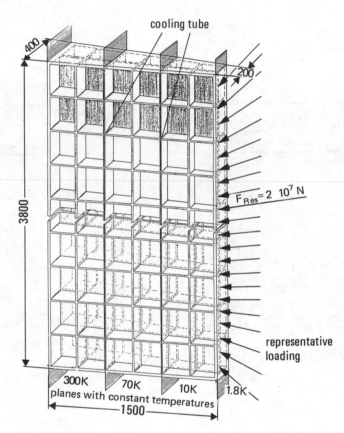

Figure 7.   Storage magnet: strut with a rib structure

The requirements on the support structure are:

(1) The heat flow must be as small as possible, because the
     required refrigeration power decisively determines the
     efficiency of the total storage unit. The support
     material should have low thermal conductivity and great
     mechanical strength. The structure should be shaped in
     such a manner that the thermal conducting cross-section
     is dimensioned exclusively by the static load and not
     with respect to additional instabilities such as buckling.
     As far as is reasonably possible, shear stresses have to
     be avoided because in comparison with normal stresses they
     require a much larger thermal conducting cross-section.

(2) The costs for material, manufacture and assembly should be
     small. For a storage unit the material required will be
     some 10^3 tons.

(3) The material must have a low outgassing rate. The support
     structure is mounted in vacuum.

(4) Long term reliability will be required. The projected
     lifetime without fracture will be 30 years.

     a) Strut Design. The Lorentz forces imposed on the support
structure vary in amount and direction with the axial location
within the winding. According to the force distribution, 15 types
of struts of a height of 3.8 m and a length of 1.5 m between 1.8 K
and 300 K each are probably needed. Intermediate cooling levels at
10 K and 70 K were chosen. More than 15,000 struts are required for
the whole storage magnet. Glass/epoxy is the most promising
structure material in this case too.

     A strut with a rib structure is proposed and shown in Fig 7.
A representative loading is drawn in. The strut consists of a
plate and ribs of box type design on both sides. For the intermediate cooling, metal tubes are embedded into the structure. The
geometrical configuration of the ribs should be adjusted to the
loading. The non-isotropic properties of the composite have to be
adapted to the force distribution ie glassfibres should be arranged
in the direction of the loading wherever possible.

     A further proposal is shown in Fig 8. The basic elements of
the strut are hollow sections with inner reinforcing ribs, which
are wrapped together. In the cooling planes there are fins (Al or
Cu) cooled by tubes wrapped with composite material. The components
are bonded together. Large adhesive areas are necessary to transmit
the large shear forces which occur at the ends of the coils.

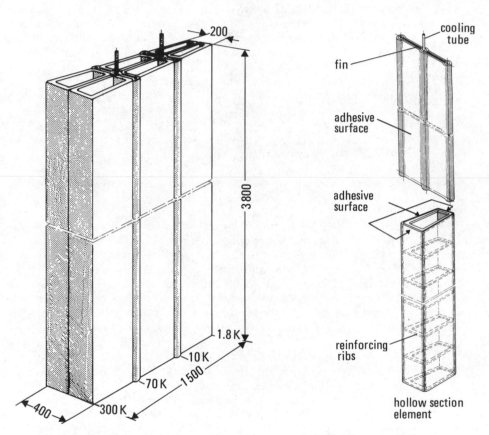

Figure 8.   Storage magnet: strut with hollow sections

b) <u>Struts and Overall Efficiency</u>. The refrigeration power of
the storage magnet is largely determined by the conduction of heat
through the support structure, which represents the main component
of the total losses. With respect to a reference unit the influence
of the struts on the efficiency is discussed.

We refer to a storage unit with a diurnal charge and discharge
cycle of 4 h working at 1 GW in each case. The estimated daily
losses of a 4 GWh-magnet are:

$$\text{total losses} \qquad W_{Lref} \;=\; 450 \text{ MWh}$$

subdivided in

$$\text{energy input of}$$
$$\text{the refrigerator} \qquad W_{Rref} \;=\; 270 \text{ MWh}$$

$$\text{other losses} \qquad W_{Oref} \;=\; 180 \text{ MWh}$$
(converter, transformers, current lead).

The thermal losses of this reference unit are based on:

- structure material glass/epoxy

- allowable compressive strength of 350 MPa

- Lorentz forces, acting under an angle of greater than 45°
  against the horizontal coil plane, were weighted by a
  factor of 2.

The efficiency of the reference unit is 90%.

For every structure, which differs in W_R (eg caused by the
allowable strength) from the reference unit, the efficiency amounts
to

$$\eta \;=\; \frac{W_{out}}{W_{in} + W_{Rref}\left(\dfrac{W_R}{W_{ref}} + \dfrac{W_{Oref}}{W_{Rref}}\right)}$$

with W_{in} - removed storage energy from the net

W_{out} - delivered storage energy to the net.

The efficiency as a function of refrigerator power is shown
in Fig 9. Based on this figure, the influence of the structure
and the properties of the material will be assessed. In comparison
pumped water storage has an efficiency of 75%.

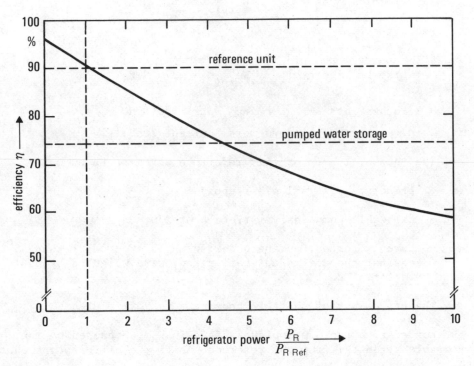

Figure 9.   Efficiency of 1GW/4GWh−superconducting storage magnet as
             as function of the refrigerator power related to a reference
             reference unit

Assuming an allowable strength of 1/5 of the reference unit (σ = 70 MPa) and an unchanged thermal conductivity of the strut material, the efficiency will only be 72%. Conversely, with the full realisable strength (350 MPa) it is increased to 94%. The influence of the mechanical and thermal properties of the structure on the efficiency of the storage magnet is clear.

CONCLUSION

Glass/epoxy is a practicable structural material in large technical devices at low temperatures. For the proposed large inductive storage systems the known properties of glass/epoxy limit the overall efficiency of the unit. A systematic material development programme, including longtime tests, will be required to achieve low thermal conductivity, high mechanical strength and low specific costs simultaneously.

ACKNOWLEDGEMENTS

I would like to thank H Kuckuck and K Meier, Siemens AG, for valuable discussions.

REFERENCES

1. W Fuchs, D Kullmann – Siemens: private communication.
2. W Amenda, P Krüger, H Lohert, K H Schmitter, M Süll, "Testing of a 0,8 m free Bore NbTi Prototype Magnet", Proceedings of the 7th Symposium on Engineering Problems of Fusion Research, Knoxville, USA, Oct. 25-28, 1977, pp. 323-326.
3. C Albrecht, "Experiences with Superconducting Magnets for Levitating the 17 tons Test Carrier EET at 150 km/h", Proceedings of the 6th International Conference on Magnet Technology (MT-6), Bratislava, Czechoslovakia, August 29 – September 2, 1977, pp. 177-188.
4. J R Heim, "A low heat leak temperature stabilized column type support", Cryogenics, Vol. 12 (1972) pp. 444-450.
5. R W Boom et al, "Wisconsin Superconductive Energy Storage Project", Vol. I (1974), Vol. II (1976), Annual Report (1977), University of Wisconsin, Madison, Wisconsin.

MACHINED GRP LAMINATES FOR COMPONENTS IN HEAVY ELECTRICAL

ENGINEERING AND THEIR USE AT VERY LOW TEMPERATURES

Heinz Fuchs

Micafil Ltd
Insulating Materials Department
CH-8048 Zurich, Switzerland

INTRODUCTION

The size of components produced from machined GRP laminates
varies from a few grams up to several tens of kilograms (Figs 1 and
2).

In practice these components are subject to complex stress
systems, often in all three dimensions. Problems of stress transfer
and differing thermal expansion coefficients are solved by correctly
adapting the shape of the component and its place in the overall
assembly to the particular anisotropic characteristics of the
material. Two properties must be known for strength calculations:

- the stress-strain behaviour, ie the relationship between
  the load and the resulting deformation

- the stresses and strains at which damage begins to occur
  in the material.

These data should be available for various temperatures and
operating conditions. Experience has shown that the designer can
produce components which are both safe and economical.[1] The purpose
of this paper is to increase his understanding of the material.

MATRIX SYSTEM AND FIBRE REINFORCEMENT

The Vetresit* laminates described here all have a cycloaliphatic

---

*Registered Trade Mark of Micafil Ltd.

Fig 1. "Pancake", part of the insulating support structure of the
        800 MJ electromagnet of the BEBC is made up of machined
        Vetresit components. To enable the liquid helium to
        circulate, the superconducting windings are spaced apart
        and the GRP plates contain slots and holes (Photo CERN).

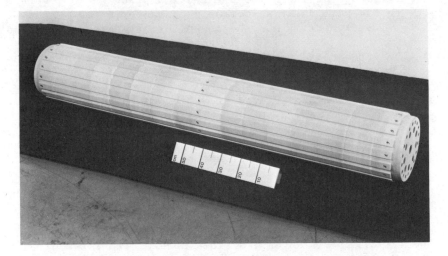

Fig 2. "Tokamak Central Pole" a bolted and glued structure of
        machined Vetresit elements (EPF Lausanne).

epoxide resin system as matrix. Although the short-term mechanical
properties of such composites are somewhat lower than laminates
based on the more common aromatic epoxide resins, they do exhibit
very good creep behaviour. In their range of application they have
low dielectric losses, are track resistant and show good resistance
to ionising radiation.

The glass content of the laminates lies between 70 and 75% by
weight (52 and 58% by volume) independent of the type of reinforcement. These highly reinforced laminates are available with the
following fibre compositions:

Vetresit	Structure, reinforcement
300	Glass mat
301	Bidirectional roving cloth (BD)
302	Roving cloth with principal reinforcement in the warp direction (UD)
311	Glass mat + BD roving cloth
312	Glass mat + UD roving cloth.

It is essential to regard the stress distribution and the
strength of the material in three dimensions, using for example,
the following convention:

Direction 1 lies in the plane of the laminate and in cloth
             reinforced laminates coincides with the warp
             direction, in mat laminates with the machine
             direction.

Direction 2 lies in the plane of the laminate and in cloth
             reinforced laminates coincides with the weft
             direction, in mat laminates with cross direction.

Direction 3 is perpendicular to the laminate plane.

It is well known that every glass fibre reinforcement has a
typical dependence of strength on angle. This anisotropy is illustrated by measurements, eg of tensile strength (Fig 3) and compressive strength (Fig 4) plotted against angle for the three principal
planes of the laminate.[2]

In cloth laminates the highest strengths are obtained in the
well-defined fibre directions. The laminates are, however, sensitive
to off-axis stresses - wherever the stress diverges by more than
10 degrees - such as occur at points of stress transfer. In

Table 1.   Tests on Laminate Vetresit 300

Property	Standard	A	B	C	Unit
Flexural strength	VSM 77103	360	430	670	N/mm^2
Impact strength	VSM 77105	160	–	250	kJ/m^2
Compressive strength – perpendicular	VSM 77102	450	400	500	N/mm^2
– parallel to the laminates	VSM 77102	270	220	100	N/mm^2
Splitting load	DIN 53463	2.8	2.2	3.5	kN

A.   Room temperature
B.   Specimens boiled for 2 hours at 100°C in water, then
      subjected to 10 cycles of cooling in liquid nitrogen and
      heating to room temperature. Test at room temperature.
C.   Test in liquid nitrogen.

considering bidirectionally reinforced laminates it should be
remembered that in practice very few components are subjected to
two equal and exactly perpendicular loads.

Design problems can arise when laminates are subjected to
interlaminar stress or when large temperature differences give rise
to stress due to the relatively large contraction perpendicular to
the plane of the laminate (Fig 5). At very low temperatures, GRP
laminates have good physical properties[3] (Table 1). However,
interlaminar strength is lowered.

The most suitable laminates for most applications are the mat
laminate Vetresit 300 and the unidirectionally reinforced mixed
laminate Vetresit 312. They were, therefore, chosen, for further
investigation. The quasi-isotropic Vetresit 300 is used for
components with multi-axial stress systems. Vetresit 312 is used
for components stressed predominantly in one direction, but where
stress transfer conditions or shear or interlaminar stress require
some transverse reinforcement.

ELASTIC PROPERTIES

To design parts which are to be stressed in several directions,
it is necessary to know the complete elastic properties. Mathema-

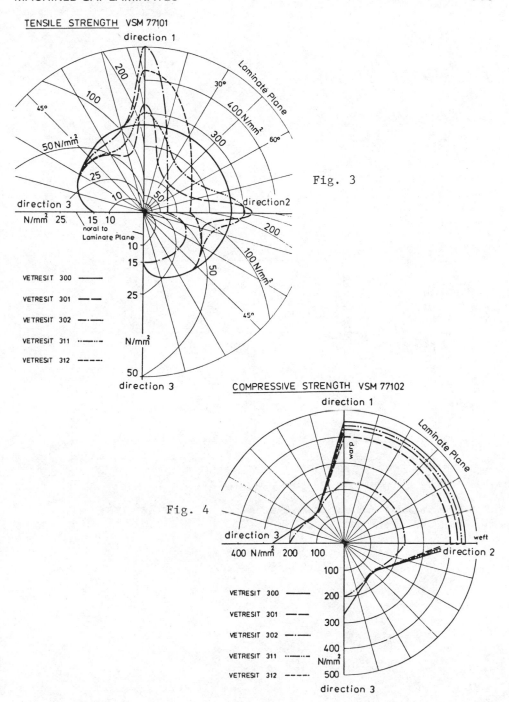

TENSILE STRENGTH VSM 77101

COMPRESSIVE STRENGTH VSM 77102

Fig. 3

Fig. 4

Figs. 3 & 4. Tensile and compressive strength of Vetresit laminates in three dimensions.

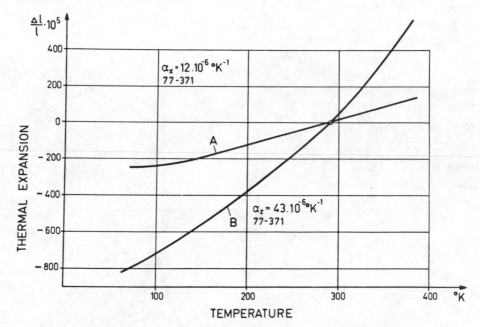

Fig 5. Thermal expansion of Vetresit 300 (mat laminate glass
        content 70 wt. %).

        Curve A in direction of laminate plane
        Curve B perpendicular to laminate plane

tically, these can best be represented as a matrix, from whose
elements the more commonly used elastic constants such as Young's
modulus and Poisson's ratio can be derived for the different
directions. If the elastic constants are referred to the principal
axes of the laminate, symmetry considerations reduce the number of
elastic constants for Vetresit 300 to 6, of which 5 are independent
of each other, and that for the less inherently symmetric Vetresit
312, to 9.   These constants can be determined from a series of
tensile and torsion tests. Experimental measurements combined with
calculations following the theories of Puck and his co-workers[5] and
Lekhnitskii[6] give the variation of the essential elastic constants
with angle in the three principal directions and planes. (For
exact details, see reference 1.) The derived results are presented
in Figs 6 and 7.

ONSET OF DAMAGE

     It is well known that in glass fibre reinforced plastics,
cracking occurs at stresses far below the fracture strength. In
cloth laminates the onset of damage is marked by a "knee" in the
stress-strain curve. In most laminates the occurrence of micro-
cracking can be detected using acoustic emission. Vetresit 300 and
Vetresit 312 were, therefore, tested in bending and tension and the
acoustic emission measured in the frequency range 10 to 50 kHz. A
conventional glass cloth laminate of type Hgw 2372 with a matrix of
aromatic epoxy resin was included for comparison.

     In tension, the cloth laminate behaved as expected (Fig 8):
a "knee" followed by an increase and a peak in the acoustic emission,
at higher strain a rapid increase in acoustic emission accompanied
by interlaminar cracking, then fracture. The stress-strain curve
for Vetresit 300 has no "knee", but is slightly curved. Except
for just before fracture, where visible interlaminar cracks occur,
no acoustic emission was detected. However, if the specimens are
repeatedly loaded and unloaded there is a narrow hysteresis loop
indicating the presence of irreversible deformation. As might be
expected from its composition, Vetresit 312 behaves as a mixture of
mat and cloth. After the "knee" there is a small increase in
acoustic emission, but this only becomes significant just before
fracture. There is no peak.

     The results of the bending tests (Fig 9), expressed in terms
of outer fibre strain, are similar. However, the "knee", the
accompanying acoustic emission and fracture itself all occur at
much higher strains. In Vetresit 312 neither "knee" nor acoustic
emission could be detected until just before fracture.

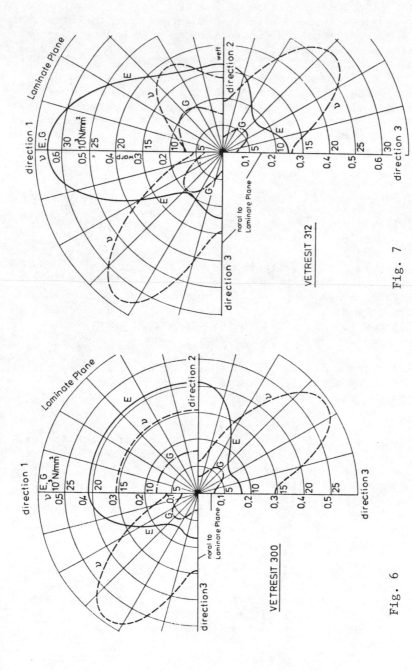

Fig. 7

Fig. 6

Figures 6 & 7. Elastic constants of Vetresit 300 and Vetresit 312; Young's modulus E, Poisson's ratio ν and shear modulus G parallel and perpendicular to the laminate plane. A three-dimensonal diagram can be built by joining the two lines marked '3.' G and ν are referred to the direction shown and to a second direction perpendicular to it in the same plane. At the axes the plane changes and with it, this second direction. This accounts for the discontinuities in G and ν.

LONG-TERM PROPERTIES

Knowledge of the long-term behaviour is essential for the design of components which are to withstand permanent loading at elevated temperatures. Creep tests at 120°C were carried out in bending on Vetresit 300 and Vetresit 312 using a 4-point bending rig with independent measurement of deflection (for exact details, see reference 7).

Initial results are given in reference 4. It became clear from subsequent measurements that plastic as well as elastic strain of the Vetresit laminates in discussion are proportional to the applied stress.

The creep strain ε for the Vetresit laminates tested is roughly proportional to the stress and increases linearly with log (time). Expressed mathematically,

$$\varepsilon = \frac{\sigma}{E} (1 + \beta \ln t)$$

where E = Young's modulus and β = constant. With t given in hours, we derive

Vetresit 300 : $\beta = 0.0347$, $E = 20800 \text{ N/mm}^2$

Vetresit 312 : $\beta = 0.0272$, $E = 35700 \text{ N/mm}^2$.

As the curves are not straight below t = 10 h, and since the properties of different batches of material differ slightly, the values of Young's modulus do not agree exactly with those given in Section "elastic properties".

Laminates based on the aromatic epoxide resin exhibit somewhat non-linear creep behaviour. The creep strain increases rather faster with log (time), particularly above 3000 h and in specimens tested at 50% of their flexural strength. Similar behaviour has been observed for unidirectional laminates stressed in various directions. On the other hand, no proportional limit or "knee" effect was observed in the creep tests.

Examples of highly stressed components made from these laminates are threaded bolts and nuts. The critical property of the bolts is the time to fracture at given load. M16 threaded bolts made of Vetresit 312 and 160 mm long were tested in tension at 120°C under static load. The force was applied to the bolts by means of nuts made of Vetresit 300 with lengths equal to twice the nominal diameter. Without jumping too rapidly to conclusions, it can be said that, after more than 65,000 hours of testing, the 2×10^5 hour creep strength will probably be more than 190 N/mm^2. The short-term strength is approximately 300 N/mm^2.

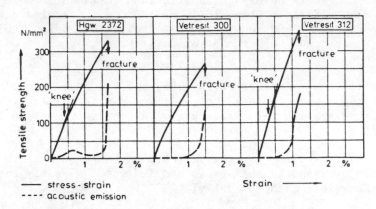

Fig. 8

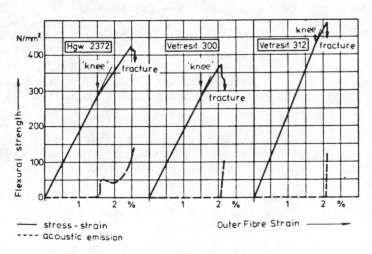

Fig. 9

Figs. 8 & 9.   Tensile and bending test.   Stress-strain curves and
acoustic emission of Vetresit 300 and Vetresit 312 in
comparison with Hgw 2372.

The mechanism responsible for creep in these materials at microscopic level cannot be detected optically. Under the microscope one sees occasional pores, or shrinkage cracks in resin-rich areas, but nothing connected definitely with creep. If the creep specimens are subsequently heated to above the glass temperature of the resin matrix, it is found that a large part of the plastic or viscoelastic strain can be recovered. In other words, creep occurs predominantly in the resin and by mechanisms that are inherently reversible; no micro-cracking, no sliding at the glass-resin interface, presumably no significant restructuring of the polymeric bonding.

Without knowing exactly which detailed mechanisms are responsible for creep, one can compare the effect of different resins. The aromatic epoxide resins used also exhibited a "memory effect", but in addition had an increased creep rate after long periods. Measurements on components made from laminates with aromatic epoxide resin suggest a similar behaviour. Experiments are running to test this hypothesis further.

Results so far show that one need have no reservation in applying the laminate studies at stresses of up to half their short-term strengths, provided creep strain and its dependence on time and temperature is taken into account. As the drop in fracture strength during the tests was only small, the risk of fracture is slight. The stresses lie below the "knee" in flexure, and the fact that there is no drop in Young's modulus confirms this. The stresses do, however, lie above the "knee" in tension. The limiting stress at which these materials can be used in practice is apparently even higher.

ACKNOWLEDGEMENT

The author thanks J H Greenwood for valuable discussions and collaboration.

REFERENCES

1.   H Fuchs and J H Greenwood, Konstruktionselemente ausbe-
          arbeiteten GFK-Presslaminaten für den Bau elektrischer
          Grossapparate, Paper 12, 13th Annual Conference AVK,
          Freudenstradt/Germany.
2.   H Fuchs, GRP for Components for the Support of the Stator
          Winding in Large Turbogenerators, MICAFIL-News, MNJ 19.
3.   H Fuchs, Insulating GFRP, Bolted Joints and their use at very
          low temperatures, 12th EEIC Bosten 1975, IEEE-Publication
          75 CH 1014-0-EI-76.
4.   H Fuchs, Aufbau and Eigenschaften von Schraubverbindungen aus
          GFK-Presslaminaten, Kunststoffe 64 (1974) 690-696.
5.   J H Greenwood, German work on GRP design, Composites 7 (1977)
          175-184.

6.   S G Lekhnitskii, Theory of elasticity of an anisotropic elastic
      body, Holden-Day, San Francisco 1963.
7.   J H Greenwood, Creep and fracture of CFRP at $180-200^{o}C$,
      Composites 6 (1975) 203-206.

MASS TRANSPORT INTO A THREE-COMPONENT COMPOSITE

SUBJECTED TO A TEMPERATURE GRADIENT

A Cervenka

Koninklijke/Shell-Laboratorium
(Shell Research B.V.)
Amsterdam, The Netherlands

INTRODUCTION

If considered on the basis of cellular structure alone, rigid polyurethane foams (PUF) consisting entirely of closed cells are ideal materials for thermal insulation applications. However, the low load-bearing capacity of the material and the vulnerability of the cell membranes make it virtually impossible for the foam to be used on its own. Therefore, technical insulations are usually based on composites consisting of (a) a substrate, which by virtue of its superior engineering properties supports the insulant (PUF), (b) A liner, which reduces ingress of species which might damage the insulant if they were to come into contact with it, and (c) the insulant itself. Examples of such composites are the insulation panels used in the building industry.

Often all three components of a composite will be pervious, so that the insulant may be permeated by species present on either side of the composite. This is particularly relevant in the case of low-temperature (cryogenic) insulations, where the atmospheric moisture can diffuse towards the cold surface and may, since the mass transport occurs along the temperature gradient, lead to phase precipitation (condensation, ice formation) in the composite.

The purpose of this paper is two-fold:

(1) to set up a theoretical model which describes the concentration profiles which develop in a three-component composite as a general function of all the relevant variables, viz component thickness, time, diffusion/ thermal properties of individual components, and the boundary conditions (concentrations, temperatures);

and

(2) to utilise the model for studying the transport of water
vapour through a three-component composite, with the aim
of assessing aspects such as when and where phase precipitation is likely to occur and the rate of penetrant
accumulation with respect to arrangement of the components.

MODEL (See Figure 1)

The composite is considered to be a slab consisting of three
components $p = 1$, 2 and 3, each of which is characterised by a
diffusion coefficient D_p, thermal conductivity λ_p and interfaces
situated at (a_{p-1}, a_p). The driving force of the mass transport is
defined to be the difference in concentrations between the interfaces a_3 and a_0; the concentration at a_3 is defined as C_0, that at
a_0 is equal to zero. The initial penetrant concentration is zero
for all arguments x, $a_0 \leq x \leq a_3$.

For a composite slab of these properties, the concentration
$c(x,t)$ at a given position x within any component p, and the time t
can be derived — using the approach of Sakai[1] — as:

$$\frac{c(x,t)}{C_0} = \frac{\displaystyle\sum_{i-1}^{p-1} \frac{a_i - a_{i-1}}{D_i} + \frac{x - a_{p-1}}{D_p}}{\displaystyle\sum_{i=1}^{3} \frac{a_i - a_{i-1}}{D_i}} + 2 \sum_{n=1}^{\infty} \frac{U_p(x,\alpha_n)}{\left.\dfrac{\delta U_3(a_3,\alpha)}{\delta\alpha}\right|_{\alpha = \alpha_n}} \times$$

$$\times \frac{\exp(-D_1\alpha_n^2 t)}{\alpha_n} \tag{1}$$

with $U_p(x,\alpha_n)$ for $p = 1$, 2 and 3 defined by the following
expressions:

$$U_1(x,\alpha) = \sin(x\alpha) \tag{2a}$$

$$U_2(x,\alpha) = (D_1/D_2)^{\frac{1}{2}} \cos\{(a_1 - a_0)\alpha\}\sin\{(x - a_1)\alpha(D_1/D_2)^{\frac{1}{2}}\}$$

$$+ \sin\{(a_1 - a_0)\alpha\}\cos\{(x - a_1)\alpha(D_1/D_2)^{\frac{1}{2}}\} \tag{2b}$$

$$U_3(x,\alpha) = \Bigg[(D_1/D_3)^{\frac{1}{2}} \cos\{(a_1 - a_0)\alpha\}\cos\{(a_2 - a_1)\alpha(D_1/D_2)^{\frac{1}{2}}\}$$

$$- (D_2/D_3)^{\frac{1}{2}} \sin\{(a_1 - a_0)\alpha\}\sin\{(a_2 - a_1)\alpha(D_1/D_2^{\frac{1}{2}}\} \Bigg]$$

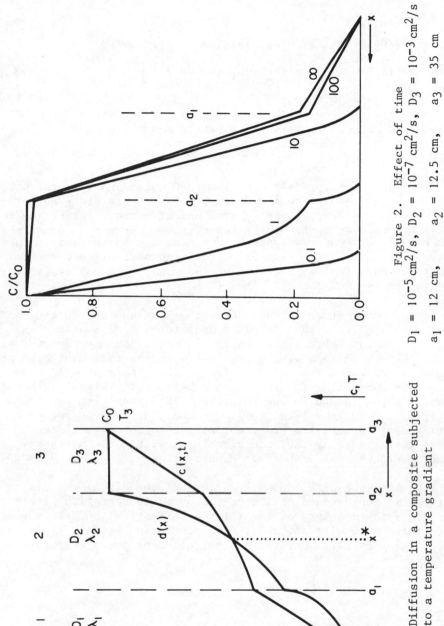

Figure 2.   Effect of time

$D_1 = 10^{-5}$ cm^2/s, $D_2 = 10^{-7}$ cm^2/s, $D_3 = 10^{-3}$ cm^2/s

$a_1 = 12$ cm, $a_2 = 12.5$ cm, $a_3 = 35$ cm

Numbers by curves indicate time (days)

Figure 1.   Diffusion in a composite subjected
to a temperature gradient

$$X \sin\{(x - a_2)\alpha(D_1/D_2)^{\frac{1}{2}}\}$$

$$+ \left[(D_1/D_2)^{\frac{1}{2}} \cos\{(a_1 - a_0)\alpha\}\sin\{(a_2 - a_1)\alpha(D_1/D_2)^{\frac{1}{2}}\} \right.$$

$$\left. + \sin\{(a_1 - a_0)\alpha\}\cos\{(a_2 - a_1)\alpha(D_1/D_2)^{\frac{1}{2}}\} \right]$$

$$X \cos\{(x - a_2)\alpha(D_1/D_3)^{\frac{1}{2}}\} \qquad (2c)$$

The symbol α_n stands for an n-th positive root of

$$U_3(a_3,\alpha) = 0 \qquad (3)$$

A concentration profile describes diffusion under isothermal conditions. As insulation always implies a temperature gradient – and thus that the temperature is x-dependent – another function has to be taken into account, namely the critical curve of a penetrant, which relates its vapour pressure/concentration to the temperature (function d(x) in Figure 1). Since the transport of heat can be shown to be much faster than that of the mass, a steady temperature state may be assumed for all times t.

If the temperature dependence of the D values is ignored, and it is postulated that the highest temperature ($T = T_3$ at $x = x_3$) governs diffusion, the mass transport is modelled as a non-restrained process conforming to a simple Fick's law in the region of high x-arguments. Provided that $d(x) > c(x,t)$, the penetrant will pass through the composite, the temperature being still high enough to prevent a phase change in the interior. This mechanism concerns the region $a_3 > x > x^*$ (Figure 1). However, as soon as the condition $c(x,t) > d(x)$ is met, the penetrant concentration is no longer allowed to increase at the rate associated with the non-restrained diffusion process, and penetrant condensation will commence at x^*.

The existence of a common intercept of the c(x,t) and d(x) functions is thus an indication of the phase condensation within the composite. The abscissa of the intercept identify the likely place of penetrant accumulation, and the concentration profile itself can be used (via its first derivative) to determine the precipitation rate:

$$F = D_p \left[\frac{\delta c(x,t)}{\delta x} \right]_{x=x^*} \qquad (4)$$

SIMULATIONS

The model is first of all used to study isothermal diffusion and the response of the concentration profile to the relevant variables; this is discussed in the section below. The behaviour of a composite subjected to a temperature gradient, and the relation

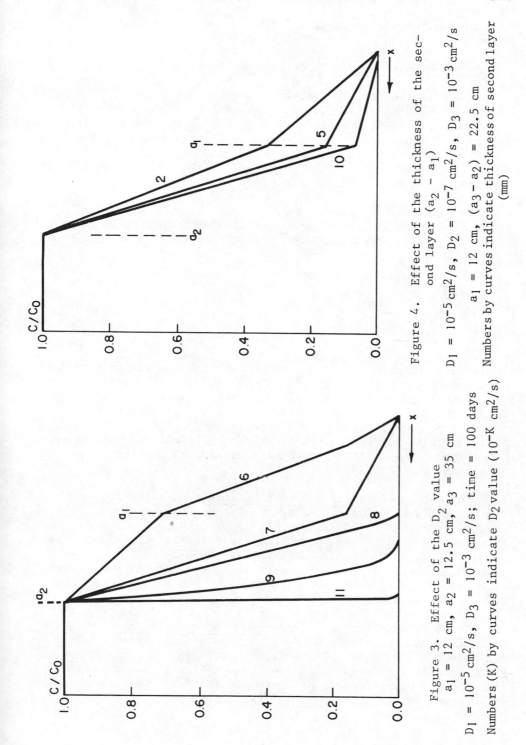

Figure 4. Effect of the thickness of the second layer $(a_2 - a_1)$

$D_1 = 10^{-5}$ cm^2/s, $D_2 = 10^{-7}$ cm^2/s, $D_3 = 10^{-3}$ cm^2/s

$a_1 = 12$ cm, $(a_3 - a_2) = 22.5$ cm

Numbers by curves indicate thickness of second layer (mm)

Figure 3. Effect of the D_2 value

$a_1 = 12$ cm, $a_2 = 12.5$ cm, $a_3 = 35$ cm

$D_1 = 10^{-5}$ cm^2/s, $D_3 = 10^{-3}$ cm^2/s; time = 100 days

Numbers (K) by curves indicate D_2 value $(10^{-K}$ cm^2/s)

between the functions $c(x,t)$ and $d(x)$ and the place and rate of phase condensation are discussed in the following section.

Isothermal Diffusion: The Effect of Variables

The effect of time is assessed for a three-component composite with $D_1 = 10^{-5}$ cm^2/s (the value characteristic for water and a medium-density PUF), $D_2 = 10^{-7}$ cm^2/s and $D_3 = 10^{-3}$ cm^2/s. The insulant thickness $a_1 - a_0$ is chosen as 12 cm and that of the liner as 0.5 cm; the overall thickness of the composite is taken as 35 cm. Results for the times between 0.1 and 10^8 days (the latter corresponding to the steady state) are given in Figure 2.

The initially stepwise function $\{c(a_3,0) - C_0; c(x,0) - 0\}$ is seen to be gradually shaped into the steady-state profile characterised by three linear dependence of the slopes:

$$\frac{C_0}{D_p \sum_{i-1}^{3} \dfrac{a_i - a_{i-1}}{D_i}} \tag{5}$$

At extremely short times (eg t = 0.1 day) the profile attains positive values within component p = 3 only. As the time progresses (t - 1 day), component p = 2 also becomes involved. Later (t = 10 days), the concentration profile develops within component p = 1 and, at the same time, approaches the steady state in component p = 3. The steady state throughout the whole composite is seen to have been achieved after 100 days for this particular composite/penetrant case.

The effect of the diffusion coefficient of the liner is illustrated in Figure 3 for D_1, D_3 and component thicknesses as above at a fixed time t = 100 days. In this way the ability of 0.5 cm thick liners of different power to retard penetrant transport is simulated. Whilst a liner characterised by a D_2 value ten times lower than the D_1 value is not very efficient for any substantial decrease in $c(x,t)$ within component p = 1, the opposite is true if $D_2 = 10^{-11}$ cm^2/s may be considered.

The influence of the thickness of the middle component on the diffusion process at t = 100 days through a composite $D_1 : D_2 : D_3 = 1 : 10^{-2} : 100$ ($D_1 = 10^{-5}$ cm^2/s) and $a_1 = a_0 = 12$ cm, as $a_3 - a_2 = 22.5$ cm, is evaluated for $l \geq 0.2$ cm in Figure 4. As expected, increasing liner thickness modulates concentration profiles $c(x,t)$ in the same way as decreasing D_2 value.

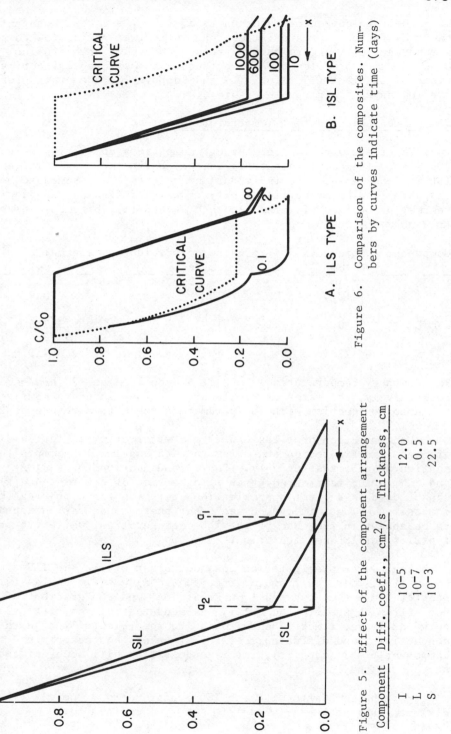

Figure 6. Comparison of the composites. Numbers by curves indicate time (days)

Figure 5. Effect of the component arrangement

Component	Diff. coeff., cm²/s	Thickness, cm
I	10^{-5}	12.0
L	10^{-7}	0.5
S	10^{-3}	22.5

When the component thicknesses $a_p = a_{p-1}$ are kept constant and their diffusion characteristics kept in a fixed ratio, the composite arrangement is found to have a pronounced effect on diffusion. Concentration profiles for three (out of a possible six) configurations of insulant (I), liner (L) and support (S), generated for $t = 100$ days, are given in Figure 5.

Composite Subjected to a Temperature Gradient

The transport of water through composites consisting of support (S), liner (L) and insulant (I) in the configurations of ILS and ISL is considered. The individual components are assumed to be characterised by properties typifying water diffusion behaviour/thermal properties of materials such as PUF (I), bitumen sealant (L) and concrete (S), eg:

Component	Thickness, cm	Diffusion coefficient, cm^2/s	Thermal conditions, $W/(m.K)$
I	12.0	10^{-5}	0.029
L	0.5	10^{-7}	0.346
S	22.5	10^{-3}	0.750

The boundary temperatures T_0 and T_3 are defined as $+20$ and $-40°C$, respectively; the water critical curve is postulated to conform to the Antoine equation with the parameters taken from reference 2.

Concentration profiles $c(x,t)$ and water critical curves $d(x)$ are shown in Figures 6A and B for the ILS and the ISL composite configurations, respectively. Phase condensation for $a_3 > x > a_0$ and $\infty > t > 0$, manifested by an intersection of $c(x,t)$ and $d(x)$, can be used as a first approximation criterion for comparing the composites. Thus, it can be seen that phase condensation appears to be absent in the case of the ILS configuration, while its presence typifies the ISL configuration.

That water precipitation does not occur within the ILS type composite is derived from the fact that $d(x) > c(x,t)$ holds for shorter times (see curve 0.1 in Figure 6A) and the gradient of the concentration profile $(\delta c/\delta x)_{x-a_3}$ exceeds that of the critical curve at longer times. Hence, owing to the relatively high thermal conductivity and diffusion coefficient values characterising component $p = 3$ (support), water condensation will occur at the surface of the composite.

For the composite with the ISL configuration, the $c(x,t)$ and $d(x)$ tangents at $x = a_3$ always $(\infty > t > 0)$ conform to:

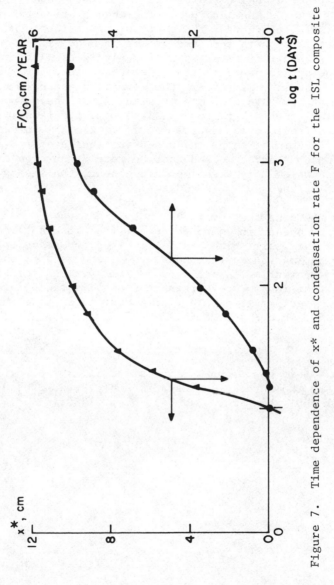

Figure 7. Time dependence of x* and condensation rate F for the ISL composite

$$\lim_{x \to a_3} \left[\frac{\delta c}{\delta x} \right] < \lim_{x \to a_3} \left[\frac{\delta d}{\delta x} \right] \tag{6}$$

Since at shorter times the concentration profiles are situated below the critical curve and $c(x,t) > d(x)$ holds for longer times (see Figure 6B), phase change and penetrant accumulation take place inside the composite. In Figure 7 the position and rate of phase condensation is correlated with the time of occurrence. It is evident that a certain "incubation" time (approximately 12 days) is necessary for the concentration profile to develop to the point where phase condensation begins. Initially the condensed phase is relatively "deep" (ie at low values of the argument x) in the composite and its accumulation rate F is low. As the time progresses, the condensed phase front moves towards the warm ($x = a_3$) interface and the rate of accumulation of the penetrant increases. Eventually, the coordinate x* and the condensation rate F converge at a certain value, after which they are time-independent, the steady state having been reached.

It can be seen that over the whole time span ($\infty > t > 0$) precipitation occurs in component $p = 1$ (the insulant) and that practically the whole thickness is involved. Moreover, the maximum rate of precipitation is reached at the steady state; when the boundary concentration $C_0 = 3 \times 10^{-3}$ g/cm^3 (value typifying solubility of water in some polyurethane foams) is taken into account, the maximum precipitation rate will amount to a rather high value of about 1.5 g water/(year.cm^2).

REFERENCES

1. S Sakai, Sci. Rep. Tohoku University, II (1922), p. 351.
2. Shuzo Ohe, Computer Aided Data Book of Vapour Pressure, Data Book Publishing Co, Tokyo, 1976, p. 1932.

FUNDAMENTALS OF NONMETALLIC MATERIALS AT LOW TEMPERATURES:

PANEL DISCUSSION SUMMARY

R L Kolek

Westinghouse Research and Development Centre
Pittsburgh, PA 15235, U.S.A.

The purpose of the panel discussion was to review large scale
applications of structural components which would be suitable for
fabrication from nonmetallic materials and composites. The author
was the moderator of the panel which consisted of the following
members:

G. Hartwig	Karlsruhe Nuclear Research Centre, FRG
A. Nyilas	KfK, Karlsruhe, FRG
W. Elsel	Siemens AG, Erlangen, FRG
A. Cervenka	Shell Chemical, Amsterdam, Holland
R. Schmid	Ciba-Geigy AG, Basel, Suisse
J. Schmidtchen	BASF, Ludwigshafen, FRG
W. Wicke	Ruhrchemie, Oberhausen, FRG
Mr. Hesselt	Messerschmitt-Bolkow-Blohm,Ottobrunn, FRG
H. Hacker	Siemens AG, Erlangen, FRG
R. Randolph	Hercules Inc., Salt Lake City, Utah, U.S.A.

Large scale projects in cryogenic technology are:

- large superconducting magnets
- large superconducting sotrage rings (see paper by W. Elsel)
- cryogenic motors and generators (see paper by H. Fuchs)
- fusion reactors.
- cryogenic wind tunnel

For all these projects there is a common problem of support
structures at cryogenic temperatures. Several components of such
structures should be made out of nonmetallic materials taking
advantage of their favourable properties relative to metals.

The discussion commenced with presentations by various panel

377

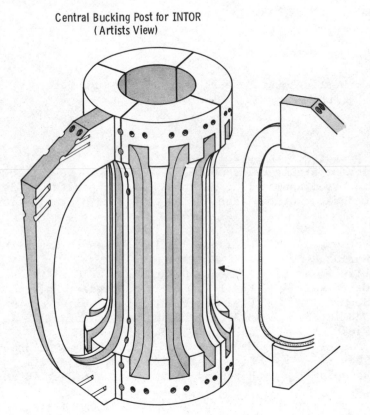

Central Bucking Post for INTOR
(Artists View)

Fig. 1.   Predesign of a Tokamak (INTOR)   (artist's view)

members which set the tone for subsequent debate. The most interesting candidates for large scale low temperature technology are
fusion reactors of the Tokamak and Mirror type. A. Nyilas
described the analysis of the Tokamak and Mirror systems and those
structures which would benefit from the properties of nonmetallic
composite materials. Both systems and the arrangement of superconducting magnets are schematically presented in Figures 1 and 2.
The specific components of the Tokamak design identified for non-
metallics were the central bucking post coil supports and the
containment case. In addition, for the Mirror design the superconducting coil support structures were also identified. A. Nyilas
made the point that current designs using metals were extremely
heavy and were fracture sensitive at the operating temperatures
required. In addition design guides did not exist for such high
loads transmitted from magnetic fields up to 12 Tesla. The most
important material requirements for large scale cryogenic support
structures are:

> high mechanical strength and stiffness at a low weight (high
> specific strength and modulus),

> low thermal conductivity and high fracture toughness.

In addition, for support structures of superconducting magnets, a
low electrical conductivity and nonmagnetic behaviour is necessary.
These requirements can be met by polymeric composites except the
fracture toughness. Polymers which are used as a matrix material
for composites show poor fracture toughness at 4 K. G. Hartwig
pointed out, that only a few polymers, e.g. high molecular weight
Polyethylene show a strain capacity of the order of 3-5% whereas
Epoxide resins do not exceed 2-2.5% failure strain at 4 K. In
addition the thermal contraction when cooled to 4 K must be
considered. If the composite fibres exhibit a low integrated
contraction (e.g. carbon fibres) then the polymeric matrix is
subjected to a pre-strain of 1-1.5%, which must be considered with
the above strain values. A detailed analysis shows that with an
epoxide resin matrix the strain capacity of Kevlar and fibre glass
composites is strongly limited by the polymer matrix. For carbon
fibre composites the situation of fibres and matrix is roughly
balanced.

The discussion also highlighted the problem of adding compounds,
to an epoxide resin formulation, to promote some flexibility at low
temperatures. The outcome of such a practice was invariably to
produce a low modulus (rubber-like) material that was prone to
excessive creep at room temperature even when employed as the matrix
material in a composite.

R. Schmid described the chemistry and types of current thermosetting matrix materials showing the best promise for low temp-

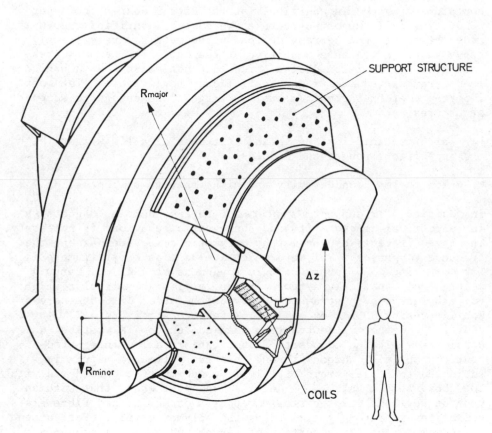

Figure 2.   End plug of a tandem mirror fusion facility

erature service.   These consisted of cycloaliphatic and bis-phenol
type epoxides as well as epoxide polyester copolymers.   Thermosetting materials such as epoxide resins are easy to handle relative
to thermoplasts like PE, but the latter have higher strain
capacities at 4 K.   The problem therefore becomes whether to improve
the low temperature properties of thermosetting materials or to
improve the fabrication and processing conditions of promising
thermoplasts.   This was a topic subsequent discussion.

R.E. Randolph and the author made a brief presentation
illustrating the variety of geometries and functional types of
composite structures currently being manufactured.   The author
specified the main technological processes for the composite
material production as follows:

    hand lay up
    bag molding
    autoclave molding
    vacuum injection molding
    premix molding
    transfer molding
    mat and preform molding
    pre preg. molding
    wet fabric molding
    filament winding
    pultrusion

It was attempted to define the feasibility of composite
materials for the fusion technology structures and those composite
matrix materials with the necessary dimensions and strain capability
at 4 K.   A concensus was quickly reached on the feasibility of
manufacturing the structures described by A. Nyilas.   It was felt
that methods to accomplish this were now available.   This was not
the case when it came to defining suitable materials.   The
situation with HDPE as explained by W. Wicke was getting enough
reinforcement into the polymer to produce a structural composite.
The high viscosity and processing characteristics of HDPE would
presently limit the fibre content to 30-40% by weight.   This
relatively low fibre content would result in a material with poor
creep performance as noted by A. Cervenka.   Since the merits of
HDPE are good strain capability at 4 K, this property served to
shape the panel's discussion of other potential polymeric materials
for fusion technology support structures.   Based on the limitations
of HDPE pointed out by W. Wicke and J. Schmidchen, the point was
made that further processing investigations might be in order to
increase the fibre content (e.g. with Fe-whiskers) and determine
creep performance at both room temperature and 4 K.   In lieu of PE,
perhaps other thermoplastic polymers should be identified which would
not have processing limitations.   As the discussion progressed, a
clear recognition of the nonmetallic material constraints for

fusion technology structures began to emerge. The free interchange
and supportive nature of the panel members which developed was
primarily responsible for the success of this portion of the discussion. The end result was the conclusion that sufficient material
knowledge does not currently exist to identify or design polymeric
composite matrix materials with the necessary long term performance
capabilities. Since HDPE has already been identified, composite
processing and testing studies are required to define its utilisation
level. Failing these approaches, new polymer materials may have to
be synthesized. The selection of materials should also include the
problem of radiation resistance.

The general feeling at the conclusion of the panel discussion
was a recognition by all members of the need for greater nonmetallic
material research for fusion energy systems. The practicality of
achieving this to be determined by a more quantitative definition
of application requirements and material useage levels.

CONFERENCE SUMMARY: FUNDAMENTAL ASPECTS AND

APPLICATIONS OF NON-METALLIC MATERIALS AT LOW TEMPERATURES

J T Morgan

Rutherford & Appleton Laboratories
Chilton
Didcot, Oxon, UK.

This second conference devoted entirely to the properties of
non-metallic materials at low temperatures, was sponsored by the
International Cryogenic Materials Conference Board. It took place
in hot summer weather at the European Organisation for Nuclear
Research (CERN) in Geneva on August 4-5 1980.

It was only in comparatively recent years that superconductivity
was referred to as 'a solution looking for a problem'. One major
problem to which the application of superconductivity may offer at
least some amelioration is the ever increasing cost of energy and
the need for its conservation. This was apparent from some of the
papers presented at this conference being devoted to the use of
non-metallic materials in superconducting devices such as energy
storage magnets, power transmission and fusion reactors. The
physical size of such projects means the application of large
quantities of non-metallic composite materials, for example, about
a thousand tonnes of fibreglass/epoxide composite in a proposed
design for a low temperature energy storage magnet.

The properties of non-metallic materials which make them
necessary in such projects are their low thermal and electrical
conductivity combined with their high strength and stiffness and
low mass. However, the less futuristic uses of these materials
for such applications as transport and storage of liquefied gases,
and space projects, were not ignored.

Twenty-three papers were presented and discussed at the
conference and a further three papers were distributed but not
presented. The papers could be divided between three broad areas:-

1. The physics of polymers at low temperature.

2. Mechanical, Electrical and Thermal properties of non-
    metallic materials at low temperature.

3. Cryogenic applications of composite materials.

Physics of polymers at low temperatures embraced topics such
as sound propagation in polymethylmethacrylate down to 0.4 Kelvin
which disclosed changes in absorption occurring at about 12 K and
2 K; and differences in thermal conductivity between semicrystalline
and amorphous polymers. Also covered was the effect of fillers on
thermal conductivity, showing a reduction in thermal conductivity
at low temperature when there is an acoustic mismatch between filler
particle and matrix.

Other physics topics included determination of Grüneisen parameters from internal friction measurement, solid superleaks and
their technological significance, and the relationship between
dynamic mechanical loss at low temperature and molecular structure
of methacrylates.

Physical and technological property data of polymers and
composites were reported in papers which covered the development of
thermally shock resistant epoxide resins, the properties of carbon
fibre reinforced laminates and graphite/epoxide resin composites.
This latter material was reported to possess the lowest thermal
conductivity and highest specific stiffness of the available reinforced plastics, making it an efficient material for the structural
support of superconducting devices. Properties measured and
reported included mechanical strength, thermal expansion and
conductivity, radiation resistance, fracture and cohesive strength.

Evaluation of the properties of non-metallic materials at low
temperature, together with the experience gained from the use of
such materials in these environments, have increased confidence in
their use, and as previously mentioned, large scale projects are
proposed for the heavy electrical industry and in energy conservation fields. Nevertheless, the conference participants felt the
need for more data, particularly long term data, and expressed the
need for the setting of engineering standards and the rigorous
specification of a number of materials in composition and manufacture followed by the documentation of relevant property data at low
temperatures.

One exploratory paper was concerned entirely with this
standardisation and possible methods of specifying non-metallic
materials for cryogenic application. Complementary to this aspect
of the technology is the need for quality control during manufacture
and a paper reviewing recent European work on the non-destructive
testing of composite materials was very useful in this context.

Epoxide resins in the form of composites appear to be the most commonly used non-metallic material for cryogenic use. From a comparison of a mat and unidirectional glass composite with conventional glass fabric laminates it appears that the former material provides a better resistance to off-axis stress, better electrical properties and improved machinability, and is capable of being produced having a high fibre content of up to 75 per cent by weight. It was pointed out however that the only materials retaining some ductility at very low temperatures were the semi-crystalline linear polymers such as polyolefins and fluorecarbons.

The conference ended with a lively Panel Discussion entitled 'The production of non-metallic materials and their application to low temperature technology'. A detailed summary is included else-where in this Volume so it is not further considered here.

Summing up, vast scope exists for non-metallic materials, they have an important contribution to make to the development of low temperature technology and especially to future energy systems. Although, at present, there are gaps in our knowledge of the properties of these materials, particularly long term performance at low temperature, hopefully this conference has provided some of the answers and highlighted some of the paths for future investigation.

We may look forward with interest to any future conferences there may be in this series.

CONTRIBUTOR INDEX

SUBJECT INDEX

Absorption, infrared, by
    polymers, 11-23
Acoustic absorption, 37, 57
Acoustic experiments, 37-47,
    49-57, 232-240, 252,
    254, 361-364
Acoustic fatigue, 306
Acoustic flaw detector, 239
Acoustic impedance, 182
Acoustic mismatch, 15, 30-32,
    182-190, 384
Activation energy, 102, 105,
    175
Adhesive bond defects, 235
Adhesives, 151
Ageing, 108, 177
Aerospace applications of
    composites, 293,
    306-325
Amorphous polymers, 1-3,
    160-180, 384
Anisotropy, and mechanical
    properties, 75, 125,
    251, 252, 262,
    357-365
Anisotropy, and superleaks,
    281, 282
Anisotrophy, and thermal
    properties, 3-23, 29,
    190, 251, 252
Antioxidants, 117-23
Applications of superleaks,
    286-290
Arrest line, 129-138
Atactic polymers, 93

Azo-bis-isobutyronitrile, 89

Bolt fracture, G.R.P., 262-264
Bond strength, 200, 216-219, 239
Boundary resistance, 30
Brillouin scattering, 49-51
Brittleness, 73, 78, 272
Bulk compressibility, 21

Capacitance thermometry, 121-123
Carbon - carbon composites,
    190-194
Carbon fibre reinforced epoxide
    resins, 186-194, 261-274,
    293-309
Carbon to carbon bond, 167
Charpy impact test, 259-274
Cocatron (servo-valve for super-
    fluid helium), 288, 289
Cohesive strength, 163, 167-177,
    384
Compact tension testing, 127-129
Compliance, 29, 215-229, 240
Composite material production
    techniques, 299-309,
    311-325, 381
Compressibility, 38
Conformation of poly (methacry-
    lates), 90-113
Cracking, 127-138, 144-147,
    151-163, 167, 209, 210,
    215-229, 239, 306, 361
Crack initiation, 127-134,
    144-147, 219, 239

389

MATERIAL INDEX

Adhesives (see also epoxide
        resins), 151, 152
Aliphatic amine curing agent
        (see also hardeners)
        158, 216
Aliphatic epoxide resins –
        see cycloaliphatic
        epoxide resins
Aliphatic polyester, 177
Alumina (aluminium oxide),
        183, 185, 193, 248
Amine antioxides, 117
Amine curing agent (see also
        Hardeners), 63, 65,
        67, 78, 158, 216
Amorphous polymers, 1-3,
        160-180, 384
Anhydride curing agents (see
        also Hardeners), 65,
        78, 81, 168
Antioxidants, 117-123
APF, 68
Aramid fibres, 206, 306, 309,
        376
Aromatic amine curing agent,
        63, 65, 67, 78
AS graphite fibres (see
        epoxide – graphite
        composites), 316
Azo-bis-isobutyronitrile, 89

Ballotini, 185, 193
BASF Polyethylene 3428, 123-138
BASF Polyethylene 3429, 123-138
BHT antioxidant (Ionol),
        118-123

Bisphenol A resins (see also
        epoxide resins), 78, 147,
        152, 156
Bitumen sealant, 374
Boron-free glass, 65
BR 14, 192, 194
n-Butyl methacrylate copolymers,
        108-113

Carbon – carbon composites
        (carbon fibres in
        pyrolitic carbon matrix),
        190-194
Carbon fibres, 186-194, 376, 384
Carborundum, 183, 185, 193
Carboxyl-terminated butadiene/
        acrylonitrile copolymer,
        81
CFRP, 186-194, 231, 232, 239,
        261-274, 293-309, 376,
        384
Chopped fibres, 199-210, 247
Ciba-Geigy 0500, 67
Ciba-Geigy 6010, 67
Ciba-Geigy CY221, 123-138, 293-306
Ciba-Geigy HY219, 82
Ciba-Geigy HY905, 123-138
Ciba-Geigy HY914, 293-306
Ciba-Geigy HY956, 123-138
Ciba-Geigy HY979, 123-138, 293-306
Ciba-Geigy LY556, 293-306
Ciba-Geigy MY790, 123-138
Ciba-Geigy X186, 123-138
Cycloaliphatic epoxide resins,
        355, 381
Cyclodecane, 106